AF768615

inics.com

GASTROENTEROLOGY CLINICS OF NORTH AMERICA

Probiotics, Prebiotics and Commensal Bacteria: Perspectives and Clinical Applications in Gastroenterology

GUEST EDITOR

Gerald Friedman, MD, PhD, MS, FACP, MACG

September 2005 • Volume 34 • Number 3

SAUNDERS

An Imprint of Elsevier, Inc.

PHILADELPHIA LONDON TORONTO MONTREAL SYDNEY TOKYO

W.B. SAUNDERS COMPANY
A Division of Elsevier Inc.

Elsevier Inc. • 1600 John F. Kennedy Blvd., Suite 1800 • Philadelphia, Pennsylvania 19103-2899

http://www.theclinics.com

GASTROENTEROLOGY CLINICS OF NORTH AMERICA
September 2005
Editor: Kerry Holland

Volume 34, Number 3
ISSN 0889-8553
ISBN 1-4160-2784-X

Gastroenterology Clinics of North America (ISSN 0889-8553) is published quarterly by W.B. Saunders Company. Corporate and Editorial Offices: Elsevier Inc., 1600 John F. Kennedy Blvd., Suite 1800, Philadelphia, PA 19103-2899. Accounting and Circulation Offices: 6277 Sea Harbor Drive, Orlando, FL 32887-4800. Periodicals postage paid at Orlando, FL 32862, and additional mailing offices. Subscription prices are $190.00 per year (US individuals), $95.00 per year (US students), $276.00 per year (US institutions), $206.00 per year (Canadian individuals), $333.00 per year (Canadian institutions), $245.00 per year (international individuals), $123.00 per year (international students), and $333.00 per year (international institutions). Foreign air speed delivery is included in all *Clinics* subscription prices. All prices are subject to change without notice. POSTMASTER: Send address changes to *Gastroenterology Clinics of North America*, W.B. Saunders Company, Periodicals Fulfillment, Orlando, FL 32887-4800. **Customer Service: 1-800-654-2452 (US). From outside of the US, call 1-407-345-4000. E-mail: hhspcs@harcourt.com**

Gastroenterology Clinics of North America is also published in Italian by II Pensiero Scientifico Editore, Rome, Italy; and in Portuguese by Interlivros Edicoes Ltda., Rua Commandante Coelho 1085, 21250 Cordovil, Rio de Janeiro, Brazil.

Gastroenterology Clinics of North America is covered in *Index Medicus, Excerpta Medica, Current Contents/ Clinical Medicine, Science Citation Index, ISI/BIOMED*, and *BIOSIS*.

Printed in the United States of America.

ELSEVIER
SAUNDERS

GASTROENTEROLOGY CLINICS
OF NORTH AMERICA

Probiotics, Prebiotics, and Commensal Bacteria: Perspectives and Clinical Applications in Gastroenterology

GUEST EDITOR

GERALD FRIEDMAN, MD, PhD, MS, FACP, MACG, Clinical Professor of Medicine, Division of Gastroenterology, Department of Medicine, The Mount Sinai School of Medicine, New York, New York

CONTRIBUTORS

STIG BENGMARK, MD, PhD, Honorary Visiting Professor, Departments of Hepatology and Surgery, Institute of Hepatology, University College, London Medical School, London, United Kingdom; and Emeritus Professor, Lund University, Lund, Sweden

PAUL BERNASCONI, MD, Director of Medical Affairs, Biocodex, Gentilly, France

JEAN-PAUL BUTS, MD, PhD, Professor of Pediatrics, Chief of the Unit of Pediatric Research, Faculty of Medicine, Cliniques Universitaires St-Luc, Catholic University of Louvain, Brussels, Belgium

MASSIMO CAMPIERI, MD, Associate Professor of Medicine, Department of Internal Medicine and Gastroenterology, Policlinic Sant'Orsola, University of Bologna, Bologna, Italy

SHIRA DORON, MD, Attending Physician, Division of Geographic Medicine and Infectious Diseases, Department of Medicine; Assistant Professor of Medicine, Tufts–New England Medical Center, Boston, Massachusetts

RICHARD N. FEDORAK, MD, FRCPC, Professor and Director, Department of Medicine, Division of Gastroenterology, University of Alberta, Edmonton, Alberta, Canada

MARTIN H. FLOCH, MS, MD, Director of Ambulatory Gastroenterology Services, Yale University School of Medicine, New Haven, Connecticut

PAOLO GIONCHETTI, MD, Assistant Professor of Medicine, Department of Internal Medicine and Gastroenterology, Policlinic Sant'Orsola, University of Bologna, Bologna, Italy

SHERWOOD L. GORBACH, MD, Attending Physician, Division of Geographic Medicine and Infectious Diseases, Department of Medicine; Director, Nutrition/Infection Unit, Tufts University School of Medicine, Boston, Massachusetts

JOHN N. GORDON, MB, ChB, Clinical Research Fellow, Division of Infection, Inflammation and Repair, University of Southampton School of Medicine, Southampton, Hampshire, United Kingdom

ERIKA ISOLAURI, MD, PhD, Professor, Department of Pediatrics, University of Turku, Turku, Finland

MARKO A. KALLIOMÄKI, MD, PhD, Department of Pediatrics, Turku University Hospital and University of Turku, Turku, Finland

KAREN M. LAMMERS, PhD, Researcher, Department of Internal Medicine and Gastroenterology, Policlinic Sant'Orsola, University of Bologna, Bologna, Italy

THOMAS T. MACDONALD, PhD, FRCPath, FMedSci, Professor of Immunology; Head, Division of Infection, Inflammation and Repair, University of Southampton School of Medicine, Southampton, Hampshire, United Kingdom

KAREN L. MADSEN, PhD, Associate Professor, Department of Medicine, Division of Gastroenterology, University of Alberta, Edmonton, Alberta, Canada

DAVID C. MONTROSE, BA, Doctoral Candidate, Center for Molecular Medicine, University of Connecticut Health Center, Farmington, Connecticut

EAMONN M.M. QUIGLEY, MD, FRCP, FACP, FACG, FRCPI, Professor of Medicine and Human Physiology, Head of the Medical School, Cork Alimentary Pharmabiotic Center, National University of Ireland, Cork, Ireland

KEVIN P. RIOUX, MD, PhD, FRCPC, Post-Doctoral Fellow, Department of Medicine, Division of Gastroenterology, University of Alberta, Edmonton, Alberta, Canada

FERNANDO RIZZELLO, MD, Assistant Professor of Medicine, Department of Internal Medicine and Gastroenterology, Policlinic Sant'Orsola, University of Bologna, Bologna, Italy

SEPPO SALMINEN, PhD, Professor, Department of Biochemistry and Food Chemistry and Functional Foods Forum, University of Turku, Turku, Finland

DAVID R. SNYDMAN, MD, Chief, Division of Geographic Medicine and Infectious Diseases, Department of Medicine; Professor of Medicine, Tufts–New England Medical Center, Boston, Massachusetts

GERALD W. TANNOCK, PhD, Department of Microbiology and Immunology, University of Otago, Dunedin, New Zealand

JON A. VANDERHOOF, MD, Chief, Section of Pediatric Gastroenterology and Nutrition University of Nebraska Medical Center, Omaha, Nebraska; Vice President, Global Medical Affairs, Mead Johnson Nutritionals, Evansville, Indiana

W. ALLAN WALKER, MD, Mucosal Immunology Laboratory, Massachusetts General Hospital and Harvard Medical School, Charlestown, Massachusetts

ROSEMARY J. YOUNG, MS, RN, Gastroenterology Clinical Nurse Specialist, University of Nebraska Medical Center, Omaha, Nebraska

ELSEVIER
SAUNDERS

GASTROENTEROLOGY CLINICS
OF NORTH AMERICA

Probiotics, Prebiotics, and Commensal Bacteria: Perspectives and Clinical Applications in Gastroenterology

CONTENTS VOLUME 34 • NUMBER 3 • SEPTEMBER 2005

ELSEVIER
SAUNDERS

GASTROENTEROLOGY CLINICS
OF NORTH AMERICA

FORTHCOMING ISSUES

December 2005

Gastrointestinal Bleeding
Don C. Rockey, MD
Guest Editor

March 2006

Palliative Care Issues in Gastroenterology
George Dunn, MD, and Todd Baron, MD
Guest Editors

RECENT ISSUES

June 2005

Irritable Bowel Syndrome
Nicholas J. Talley, MD, PhD
Guest Editor

March 2005

Obesity and the Gastroenterologist
David A. Johnson, MD, FACP, FACG
Guest Editor

December 2004

Pancreatic Diseases: Novel Mechanisms and Management
David C. Whitcomb, MD, PhD, Randall E. Brand, MD, and
Markus M. Lerch, MD, FRCP
Guest Editors

Gastroenterol Clin N Am 34 (2005) xiii–xvi
GASTROENTEROLOGY CLINICS
OF NORTH AMERICA

ELSEVIER
SAUNDERS

PREFACE

Probiotics, Prebiotics, and Commensal Bacteria: Perspectives and Clinical Applications in Gastroenterology

Gerald Friedman, MD, PhD, MS, FACP, MACG
Guest Editor

This issue of the *Gastroenterology Clinics of North America* targets the basic science and clinical applications relating to the role of commensal bacteria in health and disease. Unresolved controversies regarding the mechanisms of action and clinical use of pre- and probiotics are being resolved by basic scientists and clinical investigators. The perspectives and potential clinical applications are offered by an international group of outstanding investigators. Each investigator's work is an expression of his analytic skills and creativity, and this issue represents a rich diversity of investigative research. Our hope is that this issue will provide a platform for understanding the many potential clinical applications in this exciting area of research.

Dysregulated immune responses in genetically susceptible hosts can result in a group of inflammatory and allergic illnesses. The research gives particular emphasis to the manipulation of bacterial flora for the betterment of human host health by the addition of probiotics and prebiotics. The induction of protective immune responses in normal hosts by commensal bacteria is fully discussed.

Dr. Tannock states that the principles of microbial ecology are essential to an understanding of the relationship between the gut microbiota and the human host. A major advance in determining culturable bacterial species involves the revelation that the ribosomal subunit RNA (16S RNA) contained regions of nucleotide base sequences that were highly conserved across the world. These were interspersed with variable V regions containing the signatures of

0889-8553/05/$ – see front matter
doi:10.1016/j.gtc.2005.05.013

gastro.theclinics.com

phylogenetic groups and species. This allowed extraction of bacterial RNA and DNA-PCR amplification, making it possible to enumerate the various phylogenetic groups of bacteria inhabiting the human gut. Dr. Tannock confirms that the bacterial community of the large bowel is well regulated and resists minor perturbations. The core or true gut microbiota may be confined to relatively few populations that provide major metabolic activities. Dr. Tannock offers a research road map for future endeavors involving gut microbiota.

Drs. Kalliomaki and Walker provide a molecular and biochemical basis for the protective physiologic processes of commensal bacteria. They clearly demonstrate the interactive relationships of innate and adaptive cellular immunity with an emphasis on microbial epithelial crosstalk. The implications of these findings with reference to Crohn's disease and ulcerative colitis are examined. Possible pathways of inflammatory effects of selected probiotics offer insight as to their mechanism of action.

Drs. MacDonald and Gordon evaluate the extent to which the products of commensal flora regulate immune responses in the gut. A key to understanding the immune response was the discovery of mammalian pattern recognition receptors, the toll-like receptors whose function is to recognize conserved structures on bacteria and viruses. Signaling through toll-like receptors affects dendritic cell function, which in turn will determine T cell differentiation and antibody responses to T-dependent antigens. These authors detail the interactions of microbial flora with dendritic cells and discuss the role of bacterial flora in regulating oral tolerance.

Dr. Bengmark offers a historical perspective of the impact of dietary evolutionary changes on the incidence and prevalence of chronic diseases throughout the world. The microbial connection emphasizes how alteration of bacterial flora may play a significant role in altering human health and in particular inflammatory diseases. He examines the role of antioxidants and complex carbohydrates on chronic illnesses. Finally, he offers an analysis of commercially available pre- and probiotics and their potential clinical application for gastrointestinal diseases.

Drs. Salminen and Isolauri introduce the topic of gut inflammation and barrier function by describing the impact of indigenous microflora on immunophysiologic regulation at an early age. The importance of host–microbe interaction is most vital in the neonatal period, when the establishment of a normal microbiota provides the host with the most substantial antigen challenge, with a strong stimulatory effect on gut-associated lymphoid tissue. Impaired gut barrier may be the explanation for an infant's proneness to allergic and infectious diseases. The authors discuss the promotion of gut barrier by probiotics and the potential for clinical application.

Dr. Vanderhoof and clinical nurse specialist R.J. Young present data relating to bacterial colonization at birth and the subsequent immunologic implications for the infant. Mechanisms of probiotic actions are elucidated, followed by validation of clinical studies relating to diarrheal and allergic illnesses in the pediatric population.

Drs. Rioux, Madsen, and Fedorak discuss the role of bacteria as inciting agents evoking inflammatory bowel disease in genetically predisposed animals and humans. They summarize the literature examining therapeutic efficacy of pre- and probiotics in human inflammatory bowel disease and experimental colitis in animal studies. Evidence classifying methodologic quality of clinical trials in Crohn's disease, ulcerative colitis, and pouchitis is offered. The authors make recommendations for future basic science and clinical research.

Drs. Doron, Snydman, and Gorbach detail historical events leading to the discovery of the most studied lactobacillus: Lactobacillus GG. Bacteriologic characterization and clinical applications are offered.

Strong evidence exists supporting the use of Lactobacillus GG for the treatment and prevention of acute diarrhea in children and antibiotic-associated diarrhea. Potential applications for the treatment and prevention of allergies, childhood respiratory infections, prevention of dental caries, irritable bowel syndrome, and a wide variety of other illnesses are discussed.

Drs. Gionchetti, Lammers, Rizzelo, and Campieri provide an analysis of the basic and clinical contributions of the probiotic VSL#3. The initial studies of this agent on pouchitis provided an impetus for clinical and laboratory investigations of probiotics worldwide. Bacteriological background of patients with inflammatory bowel disease, possible mechanisms of action of probiotics, and the rationale for employing multiple strains of probiotics as therapy are discussed. Clinical studies of VSL#3 in ulcerative colitis, pouchitis, and Crohn's disease are analyzed. Current and future research agendas are discussed.

An in-depth evaluation of the only yeast-derived probiotic, *Saccharomyces boulardii*, is provided by Drs. Buts and Bernasconi. This agent has been studied in France for over 50 years. The pharmacodynamic properties, mechanisms of action, basic laboratory studies, and clinical applications are detailed. This agent does not materially affect existing microflora, yet it has properties that inactivate bacterial toxins, inhibit toxin binding, stimulate the host immune system, and provide trophic effects on the intestinal mucosa. Treatment of antibiotic-associated diarrhea and *Clostridium difficile* colitis have demonstrated efficicacy in randomized controlled trials. Further application in traveler's diarrhea, AIDS-related diarrhea, and inflammatory bowel disease are being pursued.

Dr. Quigley explains the theoretic basis for the possible use of probiotics in functional bowel disease. Particular attention is accorded the potential modification of mucosal immune processes following postinfective irritable bowel syndrome. Further evidence of low-grade inflammation and immune activation in irritable bowel syndrome suggests a role for a bacterial dysbiosis possibly correctable with probiotics. This is clearly an area of potential study for functional bowel disease.

Dr. Floch and Mr. Montrose provide an analysis of the literature pertaining to the use of probiotics in humans. They have cataloged the clinical entities and the multiple trials on adults and children. Evidence suggests efficacy in shortening childhood and adult diarrhea, preventing and treating antibiotic-associated diarrhea, and treatment of pouchitis. Further clinical applications await randomized controlled trials.

Gerald Friedman, MD, PhD, MS, FACP, MACG
Division of Gastroenterology
Department of Medicine
The Mount Sinai School of Medicine
1751 York Avenue
New York, NY 10128, USA

E-mail address: gfmd379@aol.com

Gastroenterol Clin N Am 34 (2005) 361–382

GASTROENTEROLOGY CLINICS
OF NORTH AMERICA

New Perceptions of the Gut Microbiota: Implications for Future Research

Gerald W. Tannock, PhD

Department of Microbiology and Immunology, University of Otago, 720 Cumberland Street, PO Box 56, Dunedin, New Zealand

It has been known, from early in the history of microbiology, that the gut of people and other animals is inhabited by microbial species, mostly bacteria. Louis Pasteur expressed his views to the French Academy of Sciences in 1885, on the importance of bacteria in the digestion of food, believing that life in the absence of microbes would be impossible [1]. Until the 1960s, descriptions of the collection of normal gut inhabitants were relatively simple, and *Clostridium perfringens*, lactobacilli, enterococci, and *Escherichia coli* were considered to be the predominant bacteria in the feces of adult humans [2,3]. Hungate developed innovative techniques during the 1960s for cultivating extremely oxygen-sensitive bacteria that inhabited the proximal gut of ruminants, and demonstrated the importance of these anaerobic bacteria in the rumen fermentation, and hence the reliance of the ruminant host on microbial metabolic products for nutritional well-being [4]. Holdeman and Moore [5] modified the roll-tube cultivation methods of Hungate for use in investigations of anaerobic infections of people. Certain obligate anaerobes, it became clear, were essential to the pathogenesis of these infections as had been indicated by the pioneering work of Veillon and Prevot [6]. The formulation of selective media for the cultivation of obligate anaerobes commonly implicated in anaerobic infections, together with the use of anaerobic glove boxes, accelerated the acquisition of knowledge concerning the reservoir of these anaerobes: the normal microbiota (flora or microflora) of the human body [7,8]. Now the stage was set for two major studies of the composition of the collection of microbes present in the feces (the fecal microbiota) in relation to diet and colo–rectal cancer [9,10]. These studies gathered a wealth of information concerning the composition of the fecal microbiota of people. Using the new culture techniques, some bacterial populations were detected at levels of 10^{10}/g (wet weight) of feces, and it was estimated that as many as 400 species might be capable of life in the large bowel of humans [11]. For any given

E-mail address: gerald.tannock@stonebow.otago.ac.nz

0889-8553/05/$ – see front matter
doi:10.1016/j.gtc.2005.05.006 gastro.theclinics.com

individual, however, 30 to 40 bacterial species seemed to constitute 99% of the fecal microbiota [12].

CULTURE DEPENDENCY

Investigations of the fecal microbiota, until the 1990s, relied on the use of bacteriological culture methods and microscopic observations of Gram's-stained smears [13]. Culture-dependent analysis of the fecal microbiota was flawed somewhat, however, because not all members of the microbiota could be cultivated under laboratory conditions. Even in the 1970s, researchers had observed that the total microscopic count of bacterial cells in fecal smears was always higher than the total viable count (CFUs, colony-forming units) obtained by culture on a nonselective agar medium. It was claimed, however, that good bacteriological methods would permit the culture of 88% of the total microscopic count. This comparison was obtained by using total microscopic clump counts (aggregates of bacterial cells) rather than by counting individual bacterial cells in smears [11]. Although valid from the point of view that CFUs on agar plates have not necessarily arisen from a single bacterial cell, the comparisons gave a false sense of confidence with regard to analytical results at that time. Total bacteria microscopic counts, using the 4', 6-diamidino-2-phenylindole (DAPI) stain and epifluorescence microscopy, since have revealed average total bacterial cell counts in human feces approaching 1×10^{11} per gram (wet weight) [14]. State-of the-art bacteriological methodologies still only permit about 40% of the fecal microbiota to be cultivated on nonselective agar medium in the laboratory [14]. Thus, many bacterial cells seen in microscope smears had not been investigated. Although some of these cells are probably nonviable, it seemed that many could be viable but noncultivable because of their fastidious requirements for anaerobiosis, complex nutritional interactions that were not reproduced in vitro, nutritional overload in complex media, or slow growth rates [15]. Clearly, alternative analytical methods were required.

GIVE US THE TOOLS, AND WE WILL FINISH THE JOB

Woese [16] revealed that small ribosomal subunit RNA (16S rRNA in the case of bacteria) contained regions of nucleotide base sequence that were highly conserved across the bacterial world and that these were interspersed with variable to hypervariable regions (V regions). These V regions contained the signatures of phylogenetic groups and even species. With this knowledge in hand, new methods for the analysis of bacterial communities, such as the fecal microbiota, became available. Bacterial DNA or RNA could be extracted (in theory nucleic acid from all of the bacterial types in the sample was represented in the extracts) and polymerase chain reaction (PCR) amplification (reverse transcription-PCR in the case of RNA extracts) of the 16S rRNA gene, in part or complete, could be accomplished. Clone libraries of the 16S rRNA genes could be made, and the clones sequenced, thus producing a catalog of the bacterial constituents of the ecosystem [17]. From this sequence information, it was possible to derive DNA probes that specifically targeted variable regions of the 16S rRNA gene, making it possible to enumerate the various phylogenetic groups of bacteria inhabiting the human gut regardless of whether they could be cultured [18,19]. Making a library of hundreds of clones for every sample that needed to be investigated was logistically impossible. Even microscope counts using DNA probes is a serious undertaking and really requires an automated system for unbiased results to be obtained. Therefore, a screening method to compare the bacterial composition of samples was needed, and PCR, combined with temperature or denaturing gradient gel electrophoresis (TGGE, DGGE), filled the bill [14,20]. These and other nucleic acid-based methods used in the analysis of the fecal microbiota are summarized in Table 1 [21–23]. They are critical to successful investigations of the fecal microbiota, because they detect all bacterial species regardless of whether they have been cultivated.

UNIQUELY STABLE AND METABOLICALLY PREDICTABLE

Clone libraries of 16S rRNA gene sequences prepared from human feces revealed a considerable complexity in the composition of the fecal microbiota. Moreover, a new perspective concerning the prevalence of bacterial species was obtained: Clostridia and related gram-positive genera were among the predominant bacteria (Table 2, Boxes 1–3), a feature not easily recognized from the results of culture-based studies. The *Bacteroides–Prevotella* group, the *Eubacterium rectale–C coccoides* group, and the *C leptum* group (containing the *Faecalibacterium prausnitzii* cluster) were the predominant phylogenetic divisions of bacteria encountered in all human fecal microbiota. Monitoring the composition of the fecal microbiota of individual people by the use of PCR/TGGE or PCR/DGGE, however, showed that the detailed composition of the fecal microbiota of each person, with respect to the numerically dominant species, was unique and extremely stable over time [14,20]. Variation in the composition of the human fecal microbiota between human subjects also was detected by fluorescence in situ hybridization (FISH) [24]. Collectively, these observations indicate that the bacterial community of the large bowel is well-regulated and resists minor perturbations that could be induced by a varied daily diet or minor fluctuations in host physiological parameters. Uniqueness in composition of bacterial communities likely reflects consistent physiological and immunological idiosyncrasies of people that are controlled by the genetic constitution of the host. Monozygotic twins, for example, have PCR/DGGE profiles that are more similar to one another compared with those of unrelated subjects [26].

Although the phylogenetic compositions of individual fecal microbiota seem unique among people, the overall metabolic profile of the microbiota in terms of fermentative ability is similar. The same short-chain fatty acids in similar

Table 1
Nucleic acid-based analytical methods useful in gut microbiota research

Technique	Description	Example reference
RNA dot blots	Purpose: quantification of specific bacterial population Summary: hybridization of RNA extracted from microbiota and blotted on membrane with specific labeled (often radioactive) oligonucleotide probe and universal bacterial probe. Result given as specific population as percent of total microbiota.	19
Fluorescence in situ hybridization (FISH) and epifluorescence microscopy	Purpose: detection and quantification of bacterial cells Summary: fluorescent dye-labeled oligonucleotide probe hybridizes to ribosomal RNA sequence in cells fixed on slides with wells. Enumeration by epifluorescence microscopy.	18
Terminal restriction fragment length polymorphism (T-RFLP)	Purpose: profiling and quantifying the composition of the bacterial community Summary: fluorescent dye-labeled 16S rRNA terminal gene fragments are generated by PCR using a labeled primer and subsequent restriction digestion. Detection and measurement of fluorescence using sequencing gels produce genetic fingerprints that can be related to community composition.	21
Polymerase chain reaction combined with denaturing gradient gel electrophoresis (PCR/DGGE)	Purpose: profiling the composition of bacterial communities for comparative analysis Summary: a variable sequence region of the 16S rRNA gene is amplified by PCR from genomic sequences in the sample. Separation of 16S rDNA fragments from different bacterial types is based on differences in chemical stability, through a linearly increasing gradient of chemical denaturants. The profile of DNA fragments represents the genetic fingerprint of the community.	24
Polymerase chain reaction combined with temperature gradient gel electrophoresis (PCR/TGGE)	Like PCR/DGGE but using temperature gradient in separation of DNA fragments	20

Table 1
(continued)

Technique	Description	Example reference
16S rRNA gene library	Purpose: preparing a phylogenetic catalogue of the bacterial community Summary: PCR amplification of 16S rRNA genes from a sample, subsequent cloning in a plasmid vector and cloning host to derive a library of individual rRNA gene clones for sequencing and phylogenetic analysis	22
Real time quantitative polymerase chain reaction	Purpose: quantification of bacteria Summary: PCR primers and a labeled probe (often incorporating a reporter dye and a quencher molecule) are used to measure the real-time accumulation of a specific target sequence	23

proportions are detected in human feces regardless of the microbiota profile obtained by PCR/DGGE [27]. This argues for considerable redundancy among the bacterial species that can inhabit the gut; probably several bacterial species can fill a given ecological niche, and each niche is filled differently from person to person. Thus, no matter what the composition of the microbiota, the bowel ecosystem functions in the same, predictable manner. It may be speculated, therefore, that amassing catalogs of bacterial species will not promote a much clearer understanding of the gut ecosystem than that which already exists. Mere detection of a bacterial species in a gut sample reveals little of the role of this organism in the ecosystem, or even if it is metabolically active. There is a striking difference in PCR/DGGE profiles of bacterial communities in human feces generated using bacterial RNA as the PCR template compared with DNA. Profiles generated from bacterial RNA in a recent study showed intensely stained fragments clustered in the middle of the denaturing gradient. The profiles were markedly different than those generated from DNA [28]. RNA extracted from bacterial cells is mostly ribosomal RNA and can be used as an indicator of metabolic activity, because the ribosome per cell ratio is roughly proportional to growth rate of the bacteria. Although DNA-based analytical procedures provide a phylogenetic picture of the community, they do not reflect metabolic activity, because the DNA could originate from living active cells, living dormant cells, lysed cells, or dead cells [29]. On the basis of these observations, it can be proposed that the human gut provides a habitat for a diversity of bacterial species, as seen in DNA-DGGE profiles, the composite of which varies from one person to

Table 2
Bacterial groups quantified by oligonucleotide probes in the feces of adult humans[a]

Target bacterial group	Percentage of 6-diamidino-2-phenylindole count (total cells)
Atopobium group	7.0
***Bacteroides/Prevotella* group**[b]	**18.0**
Bifidobacterium genus	3.0
Clostridium histolyticum group	< 0.1
Clostridium lituseburense group	< 0.1
Coriobacterium subgroup	1.0
Escherichia coli	0.2
Eubacterium cylindroides group	0.9
***Eubacterium rectale-Clostridium coccoides* group**	**14.0**
***Fusobacterium prausnitzii* group**	**16.0**
Lactobacillus/Enterococcus group	< 0.1
Phascolarctobacterium	0.5
Ruminococcus subgroup	6.0
Veillonella	< 0.1

[a]After Harmsen and Welling [25].
[b]The predominant phylogenetic groups are shown in bold.

another. The core or true gut microbiota may be comprised of relatively few populations that provide the major metabolic activities.

LOCATION, LOCATION, LOCATION

Most investigations of the gut microbiota of people have focused on the bacterial content of feces. Does what is found in feces, however, reflect the composition of the gut microbiota? The answer is a qualified yes, because the gut microbiota of people is located, for the most part, in the colon. Moore and Holdeman [10] demonstrated, by the examination of samples collected from sudden death victims, that the composition of the fecal microbiota resembled that of the distal colon. The qualification must be that the metabolic activity of the microbiota differs between the proximal and distal colonic regions [30]. This is a reflection of the relative availability of fermentable substrates in the two sites. The colon represents an anaerobic chemostat, in which the digesta and host secretions provide a culture medium that is especially rich in plant polymers and glycoproteins of complex structure. The bacterial community feasts on these substrates in an interactive manner, producing a distinctive pattern, qualitatively and quantitatively, of metabolic end products. The feces represent the chemostat effluent. It could be argued, then, that microbiological information obtained from feces does not provide accurate information about the metabolic structure (biological networks) of the gut community. Interest in the role of the gut microbiota in the pathogenesis of inflammatory bowel diseases (IBD) has stimulated interest in obtaining samples from the human gut itself. IBD (Crohn's disease and ulcerative colitis), it

Box 1: Examples of cultivated bacterial species within the *Bacteroides-Prevotella* group

Porphyromonas salivosa
Porphyromonas catoniae
Porphyromonas gingivalis
Porphyromonas cangingivalis
Porphyromonas endodontalis
Rikenella microfusus
Cytophaga fermentans
Bacteroides caccae
Bacteroides ovatus
Bacteroides thetaiotaomicron
Bacteroides forsythus
Bacteroides distasonis
Bacteroides merdae
Bacteroides putredinis
Bacteroides splanchnicus
Bacteroides vulgatus
Bacteroides stercoris
Bacteroides uniformis
Bacteroides eggerthii
Bacteroides fragilis
Prevotella heparinolyticus
Prevotella pallens
Prevotella veroralis
Prevotella denticola
Prevotella oulora
Prevotella oris
Prevotella oralis
Prevotella ruminicola

seems, results from a genetically predisposed loss of tolerance by the immune system to the presence of the gut microbiota. The latter provides a constant antigenic stimulus for the host's immune system [31–33]. Normally, immunological tolerance toward the gut microbiota prevents continuous intestinal inflammation [34]. There is evidence from people and experimental animal models of IBD, that this controlled, homeostatic response is lost in genetically susceptible hosts (dysfunctional immune system), leading to the development of chronic immune-mediated inflammation of the intestinal tissue

Box 2: Examples of cultivated bacterial species within the *Eubacterium rectale-Clostridium coccoides* group

Clostridium polysaccharolyticum
Clostridium herbivorans
Clostridium populeti
Clostridium coccoides
Clostridium nexile
Clostridium oroticum
Clostridium clostridiiforme
Clostridium celerecrescens
Clostridium xylanolyticum
Clostridium symbiosum
Clostridium aminovalericum
Clostridium aminophilum
Coprococcus eutactus
Butyrivibrio crossotus
Butyrivibrio fibrisolvens
Eubacterium xylanophilum
Eubacterium ventriosum
Eubacterium eligens
Eubacterium formicigenerans
Eubacterium contortum
Eubacterium rectale
Eubacterium hadrum
Eubacterium halii
Eubacterium ramulus
Lachnospira pectinoschiza
Ruminococcus obeum
Ruminococcus hansenii
Ruminococcus productus
Ruminococcus torques
Ruminococcus gnavus
Roseburia cecicola

[31,32]. Mucosa-associated bacteria would be, because of their intimate location in relation to host tissue, the cells most likely to be encountered by the gut-associated immune system, and therefore the ones most likely to stimulate inflammatory processes.

Box 3: Examples of cultivated bacterial species within the *Clostridium leptum* group

Faecalibacterium prausnitzii
Ruminococcus bromii
Eubacterium siraeum
Eubacterium plautii
Eubacterium desmolans
Clostridium sporosphaeroides
Clostridium viride
Clostridium leptum
Ruminococcus flavefaciens
Termitobacter aceticus

Although permanent epithelial associations by nonpathogenic bacteria have been demonstrated in several animal species [35], it is unlikely that exactly the same situation pertains to people. There are differences in the anatomy and histology of the human gut relative to that of other monogastric animals. The formation of *Lactobacillus* biofilms in the gut of mice and chickens is, for example, conditional on the presence of a nonsecretory, squamous epithelium such as that of the rodent forestomach or avian crop [35]. The gut of humans lacks this prerequisite for biofilm formation, because it is lined by an epithelium composed of columnar cells. Bacteria could become trapped and possibly multiply in the mucus layer that is continuously secreted onto the mucosal surface of the human gut. Indeed, in the proximal colon of rodents, a bacterial community distinct from that of the lumen has been observed in the mucus layer [36,37]. Studies aimed at defining the composition of the human mucosa-associated (biopsy-associated) microbiota have been reported, but different investigations seem to produce contradictory results [38,39].

Accurate definition of the mucosa-associated microbiota, should one exist, will be difficult to achieve, because mucosal biopsies must be collected: an invasive procedure that is ill-defined in terms of collection procedure from study to study. Before biopsy collection, the patient is purged to remove colonic contents. The effect of this preparative treatment on the composition of the microbiota is unknown. Further, the bowel is not decontaminated completely by this procedure, and a fecal fluid continues to be present in the bowel to bathe mucosal surfaces. Therefore, it is not clear what exactly is being sampled in current studies: the mucosal surface contaminated with lumenal bacteria or true mucosa-associated inhabitants. Additionally, the extent of contamination of the colonoscope with bacteria from the fecal fluid has not been determined. Therefore, reports about the mucosa-associated microbiota of people must be treated with caution, but at least these studies have begun to relieve the previous focus on the microbiota of feces.

If bacteria truly resident in mucus or on the mucosal surface exist, it is likely that they vary from patient to patient as a reflection of the known uniqueness of fecal microbiota. Therefore, comparisons of healthy subjects and patients will not be useful. Intra-individual comparisons of the microbiota (inflamed, noninflamed mucosa from the same subject) will be useful, however, because recognition of microbiota of abnormal composition could provide a therapeutic or diagnostic target. These observations will not, however, permit a causative role in IBD to be assigned to specific bacterial populations. Alterations in the mucosa-associated microbiota could occur as a result of inflammation-induced changes in the habitat. Nevertheless, it is important to determine the spatial distribution of the microbiota in the human gut, and studies with biopsies are necessary.

WHERE DO YOU THINK YOU COME FROM?

The question of human origin probably worried humans for thousands of years. This question has probably worried humans for thousands of years, but has certainly been the case from the time of the first performance of Sophocles' play "Oedipus the King", and much developed by Sigmund Freud hundreds of years later. As a result of not knowing his immediate ancestry, Oedipus unfortunately and unknowingly married his mother and killed his father, a rather catastrophic outcome of ignorance. A sound knowledge of the origin of bacteria isolated from human feces is necessary to avoid scientific catastrophe. This is because recent studies have shown that the detection of a particular bacterial species in human feces does not necessarily mean that the bacteria are inhabitants of the gut. *Lactobacillus* populations in human feces, for example, have been observed to fluctuate in species and strain composition [14]. Most probably, these *Lactobacillus* species were transient in the gut, because they are ubiquitous in foods and often used as starter organisms in food production. Food-associated lactobacilli survive passage through the gut but do not persist there in the absence of continued consumption [14,40,41]. Their presence in the gut is dependent on external factors (consumption of food in which they are present) and can be termed allochthonous (found in a place other than where they were formed). In contrast, some *Lactobacillus* species can be detected in the feces of a given human host consistently over long periods of time. These lactobacilli, notably *L ruminis*, can be referred to as being autochthonous to the gut (found where formed; inhabiting a place or region from earliest times) [14,42].

First postulated in relation to the gut microbiota by Dubos and colleagues [42,43], autochthony recently was [44] defined as: "an autochthonous species has a long-term association with a particular host species, forming a stable population of characteristic size in a particular region of the gut, and has a demonstrable ecological function." Certain lactobacilli provide good examples of autochthonous species. In the case of broiler chickens raised under commercial conditions [45,46], lactobacilli become established in the crops (predictable, stable population in specific region) of the birds soon after hatching and persist throughout the life of the host despite the common administration of antimicrobial drugs in the poultry feed (long-term association with a particular host species). At least some *Lactobacillus* strains adhere to the crop epithelium and proliferate to form a biofilm. The metabolic activities of the lactobacilli that persist in this way influence the pH of the digesta, which, in turn, inhibits the proliferation of enterobacteria (demonstrable ecological function) [47].

Discovery of the molecular traits that confer autochthony on a bacterial species is of interest, because it will tell about the evolution of the gut microbiota and their relationship with the host. The bacteria residing in the gut of mammals have coevolved with their host and have developed a high degree of adaptation and specialization. These bacteria must possess traits that enable them to establish and maintain themselves in a lotic and highly competitive environment. The challenge for the bacteria is to satisfy their own growth requirements and cope with the hostile conditions generated by the competing members of the microbiota and by the defense mechanisms of the host. Lactobacilli that are autochthonous to the gut of rodents are providing clues in these respects as a result of the application of in vivo expression technology, mutation of specific genes, and in vivo observations of ecological performance [48,49].

COMPETITIVE EXCLUSION AND HOMEOSTASIS

Probiotics (dietary supplements or foods that contain living microbial cells that are believed to promote health) have enjoyed an uncritical popularity in recent decades [50]. Historically, a probiotic was thought to alter the composition of gut microbiota, but recent studies have shown that this does not occur beyond a transient modification of a target population. Ingestion of lactobacilli, for example, modifies the *Lactobacillus* population [14,51–54]. Representative of these studies is one conducted by Tannock and colleagues [14], in which molecular typing (genetic fingerprinting) of *Lactobacillus* isolates was used to follow the fate of a probiotic strain in the gut of humans. The probiotic strain, *L rhamnosus* DR20, could be differentiated from other lactobacilli using genetic fingerprinting. In the case of 9 out of 10 subjects, *L rhamnosus* DR20 could be detected by culture of the feces during the period of probiotic consumption. In eight of the nine subjects, the ingested *Lactobacillus* strain was no longer detectable in the feces by 1 month after consumption of the probiotic ceased. DR20 was not detected by culture in the feces of a tenth subject at any stage of the study. Administration of the probiotic did not otherwise alter the composition of the fecal microbiota, and indeed, as is commonly observed, lactobacilli, even when administered in relatively large numbers, comprised less than 1% of the total fecal microbiota [14]. Under these conditions, it is doubtful that probiotics have any impact on the large bowel ecosystem, but they might be effective in the small bowel. After all, consumption of a probiotic product delivers about 10^9 allochthonous bacterial cells to this site with every dose. A realistic probiotic target may not be the modification of the colonic microbiota, but the stimulation of the mucosal immune system of the small intestine.

Saavedra [55] has summarized studies in which lactic acid-producing bacteria have been administered to children with the aim of preventing or minimizing diarrheal diseases. He concluded that the impact of the probiotic products appeared to be most significant against rotavirus infections, and he suggested that an immunological mechanism was responsible for the beneficial effects.

Maintenance of a self-regulating community structure explains the stability (homeostasis) of gut microbiota composition. The regulating mechanisms are not known in detail but, as in nature in general, they probably involve competition, amensalism, parasitism, and predation [56]. The greatest of these as regulatory effector is doubtless competition, specifically the competitive exclusion principle, perhaps better termed the niche exclusion principle. Put simply, two types of organism (species or genotype) with a similar ecology (fill the same ecological niche) cannot live together in the same place. In other words, organisms that are in competition for the identical habitat cannot coexist [57,58]. The large bowel ecosystem is inhabited by diverse bacterial species. The biodiversity of the community is caused by the heterogeneous environments that exist in relation to bacterial nutrients and spatial arrangements. Hundreds if not thousands of niches must exist, and different types of bacteria are favored in each of them. The large bowel community, therefore, is characterized by a richness of coexisting bacterial species.

A feature of the results of microbial ecological studies of gut microbiota is not only the diversity of species that are present, but also the diversity of genotypes (strains) within a species. This has been demonstrated by comparisons of genetic fingerprints of bacterial isolates generated by restriction endonuclease digestion of bacterial genomes followed by agarose gel electrophoresis [59,60]. It is difficult to imagine how all of these strains can coexist in the gut unless they have differential phenotypes [57]. Studies to determine the genetic basis of ecological differentiation of bacterial strains, the necessary condition for coexistence, although likely to be technically difficult, should be pursued.

Modification of microbiota composition by the consumption of prebiotics (potential substrates for bacterial inhabitants of the intestine) can be attempted [61,62]. This approach has the rationale that autochthonous bacterial strains could be encouraged to proliferate in the gut, hopefully with beneficial consequences to the consumer. Oligosaccharides and inulin have been reported to selectively increase the numbers of bifidobacteria in the feces, but most of the studies have involved the consumption of prebiotic powders [63–67]. The results of these studies may not be relevant to everyday life, because the substrates, for commercial use, would be incorporated into a food product. Studies in which people consume prebiotic substances in real food would seem to be more relevant [28,68].

There remains also a conceptual problem in relation to prebiotics. Carbon and energy sources are provided to the colonic bacteria by supplementing the diet with oligosaccharide. More fermentable substrate becomes available for bacterial metabolism, but, in the highly regulated bacterial community, other nutrients would continue to be present in growth-limiting amounts. Therefore, an increase in bifidobacterial or other specific bacterial cell numbers, on the basis of knowledge of bacterial physiology, should not occur [28].

BLUE PRINTS AND DESIGN PRINCIPLES

Large-scale DNA sequencing methods have provided the ability to unravel and compile the genetic secrets of bacteria. The bacterial cell's capabilities are reflected in, and dictated by, its genomic sequence. This sequence, therefore, forms the blue print for the cell's biochemical and functional endowment. Annotation of DNA sequences and bioinformatic evaluation of predicted genes has provided ecological information about three anaerobic residents of the human gut: *Bifidobacterium longum, Bacteroides fragilis* and *Bacteroides thetaiotaomicron* [69–71]. The carbon and energy requirements of the large bowel microbiota are met from two sources: complex carbohydrates, proteins and fats that have escaped digestion by host processes in the small bowel, and the components of host secretions and sloughed epithelial cells. Probably of particular importance in these respects are plant polysaccharides such as cellulose, pectins, arabinoxylans (hemicellulose), resistant starch, β-glucans derived from the diet of the host, and large, heavily glycosylated glycoproteins (mucins) that are the principle components of mucus, which is produced abundantly by goblet cells in the intestinal mucosa [72,73]. *B longum, B fragilis,* and *B thetaiotaomicron* devote much of their genome to genes encoding hydrolytic enzymes, often contained in self-regulated operons, which catalyze the degradation of oligosaccharides and glycoproteins. The genome also encodes signal transduction mechanisms, by which the bacteria can detect and react to changes in the nutrient milieu. Such responses to qualitative changes in substrate availability, and well-regulated metabolic pathways that conserve cellular energy, doubtless help explain the ecological competitiveness of these bacteria and the stability in composition of the microbiota. Additionally, the genome of *B fragilis* NCTC9343 encodes the ability to modulate cell surface antigenicity by producing at least eight distinctly different capsular polysaccharides. This is achieved by regulating the expression of polysaccharide-encoding loci in an on–off manner by reversible inversion of DNA segments that contain the promoters for their expression [70]. The ability to modulate antigenicity of the bacterial cell surface may be a mechanism by which to evade the host immune response and hence aid bacterial persistence in the gut.

Apart from this relatively small amount of genomic knowledge, however, little information about the processes (the design principles) that shape and sustain the large bowel ecosystem is available. Metagenomics is an investigative approach by which this information could be obtained. Metagenomics is a facet of synecology in which the microbial community is considered to be a single organism with multiple chromosomes (collective genomes) rather than diverse species with individual genomes [74]. The metagenomic approach entails cloning and sequencing large fragments of community genomic DNA that have been extracted directly from the ecosystem of choice. This abrogates the problem of noncultivability of most community inhabitants. The cloned DNA fragments are large enough to encode operons and therefore might result in the expression, by a surrogate bacterial host, of several enzymes that could catalyze a relatively complex metabolic process, including the synthesis of secondary metabolites [75].

Metagenomic libraries derived from microbial community genomes can be screened for heterologous phenotypic traits that include enzymes and other proteins that are essential to the functioning ecosystem [76]. Hence, the libraries provide a means of assessing details of community biochemistry and genetics, and of accessing bioactive substances. Metagenomics holds promise for use in investigating the pathogenesis of IBD, because the important factor in the production of these diseases is the nature of the bacterial products to which the dysfunctional immune system reacts. Although most members of the gut microbiota are noncultivable, it should be possible to detect the immunoreactive substances that they produce by cloning high molecular weight DNA extracted from the gut microbiota and then screening the metagenomic library for the production of proinflammatory antigens.

SYMBIOSIS OR SOMETHING ELSE?

The relationship between the gut microbiota and the host has been described as a symbiosis [77]. Strictly speaking, symbiosis refers to the intimate coexistence of dissimilar species that exert mutually beneficial effects. Often, only two biological partners are involved. Commonly, one, and sometimes both, of the symbionts are linked in an obligate relationship, without which they are unable to exist. An important feature of symbiosis is the permanency of the relationship in which the association lasts for the major part of the life of the symbionts. An example of a symbiosis is the lichen: an association between a fungus and an alga that produces an external structure, a characteristic physiology, and an ecologically independent existence so distinct as to be considered and classified as a separate organism [56]. Microbes resident in the rumen of herbivorous animals such as sheep and cattle, and in the intestines of termites, clearly form a symbiosis with their host, because the latter is unable to digest the food, of high cellulose content, that it ingests. Put simply, in the absence of the microsymbionts in the gut, the host would starve to death, because its food would not be degraded, and microbial metabolic products upon which the host relies for carbon and energy would not be produced. Description of the relationship between the gut microbiota and monogastric mammalian hosts such as people as a symbiosis must be considered to be a rather loose use of the term. People, and other animal species, can exist independently of their microbiota (germ-free animals, including people, and pure cultures of gut bacteria have been derived) [78,79], but it is of course normal that they associate throughout most of the life of the host. The relationship benefits the microbes, because they are provided with nutrients by the host from exogenous and endogenous sources, but it is not completely clear as to the benefits accrued by the host from the association. It is true that, from the results of gnotobiotic animal experiments, the presence of the gut microbiota impacts on gut physiology [1,79–83]. In most studies, these effects are produced in the presence of a complex community that is composed of numerous bacterial species. It may be valid to consider the microbial community of the gut as a single organism, in which case an argument for a mutually beneficial association can be made. Very few bacterial species from this community have, however, been tested in monoassociated animal experiments to determine the impact of distinct groups of bacteria on the host, and only one (*B thetaiotaomicron*) has been demonstrated to confer what might be considered beneficial physiological effects on the host [80,82]. At this stage, therefore, a more correct description of the relationship between most members of the gut microbiota and the human host is that of commensalism. Because of the rather unsatisfactory state of current knowledge, individual members of the microbiota can be considered to be commensals, because they obtain benefit (a habitat) from the association, whereas the host receives neither demonstrable benefit nor harm. There is only a fine line to be drawn between commensalism and parasitism. Symbiotic relationships commonly contain an element of mutual exploitation and have even been described as covert parasitism [56]. Indeed, the microbiota–human relationship, under appropriate predisposing circumstances, can develop quickly into that of parasite host relationship, in which the host is injured by the association. Sepsis caused by obligate anaerobes or facultative anaerobes that are members of the normal microbiota provides an example [6].

IMMUNOLOGICAL FREUDIANISM

Many clues to the influences of bacteria on the animal host have been obtained from comparisons of the biochemical and physiological characteristics of gnotobiotic animals [1,79]. Gnotobiotic research can be done at a sophisticated level because of the availability of genome sequences of experimental animal species and the consequent preparation of DNA microarrays that can be used to measure gene expression. Gordon and colleagues have studied the impact of colonization of formerly germ-free mice by the bacterial species *B thetaiotaomicron* [80–82]. Perhaps most striking have been their observations concerning angiogenesis in the murine gut. Quantitative three-dimensional imaging studies showed that a plexus of branched and interconnected blood vessels developed postnatally in small bowel villi of conventional mice. Angiogenesis coincided with the establishment of the gut microbiota. Vascular development was arrested in germ-free mice, but could be restarted by colonization of the gut by a conventional gut microbiota, or by *B thetaiotaomicron* that had been shown in other experiments to up-regulate expression of the murine angiogenin-3 gene in the ileal mucosa [80,82].

The gut of newborn infants resembles that of the germ-free animal because it is not yet colonized by microbiota. This germ-free state is short-lived, however, because within minutes of birth, bacteria of maternal and environmental origins have the opportunity to colonize the neonate. Seemingly unchecked proliferation of bacteria proceeds initially in the neonatal gut, resulting in a heterogeneous collection of bacterial species. Subsequently, regulatory mechanisms generated within the ecosystem (autogenic factors) and by external forces (allogenic factors) permit the persistence of some bacterial populations, but the elimination of others in a continuous succession. It takes several years to produce a climax community similar to that of adults [84].

Members of the bacterial genus *Bifidobacterium* are numerically predominant in the gut of infants during the first months of life, a phenomenon first described in 1905 by Tissier [85]. *Bifidobacterium* species could be of particular relevance to the development of neonatal gut physiology, because they are the numerically predominant bacteria during the first months of life, regardless of diet. Using nucleic acid-based methods of analysis, bifidobacteria have been shown to form between 60% and 91% of the total bacterial community in the feces of breastfed babies and 28% to 75% (average 50%) in formula-fed infants. In the feces of adults, they comprise, on average, only a few percent of the microbiota [86]. Thus, these bacteria could have an important role in early life on the development of host characteristics.

In recent decades, many affluent countries have experienced an increase in the prevalence of atopic diseases, including asthma [87]. Several aspects of lifestyle have changed in these countries over the same period, and theories have been advanced to explain the altered prevalence of allergies. The hygiene hypothesis proposes that atopic diseases could be prevented by infections in early childhood, because the neonatal immune system would be driven toward a T helper 1 response. A specific infectious protective factor, however, has not been identified [88]. Attention has turned to the gut microbiota and the possibility that colonization of the gut by specific bacterial species may be more important than the impact of sporadic infections [88–90].

The environmental conditions under which babies are born and nurtured may affect which microbes they are exposed to, and subsequently influence the composition of their gut microbiota. Differences in neonatal gut microbiota might occur because of the common occurrence of hospital deliveries, caesarean sections, special care baby unit admissions, smaller family size, widespread use of antibiotics, good hygiene, and differences in maternal diet in affluent countries. Either the lack of exposure of babies to particular bifidobacterial species, or elimination of bifidobacterial species from the gut through the use of antibiotics, might reduce the exposure of children to important bacterial antigens at a critical time in the maturation of the immune system. Hence, the difference in national prevalence of allergies in children offers scope for investigating the possible preventative role of specific gut bacteria against atopic diseases.

A recent report describes a study in which the fecal populations of bifidobacteria from children aged 25 to 35 days in Ghana (low prevalence of atopy), New Zealand, and the United Kingdom (high-prevalence countries) were compared [91]. Natal origin influenced the detection of bifidobacterial species, because fecal samples from Ghana almost exclusively contained *Bifidobacterium infantis,* whereas those of the other children did not. Choosing species on the basis of the bacteriological results, bifidobacterial preparations were tested for their effect on cell surface markers and cytokine production by dendritic cells harvested from cord blood. Dendritic cells were used, because they are principal antigen-presenting cells that are present at mucosal surfaces, including that of the gut, and are therefore likely to have an important role in the interplay occurring between the gut microbiota and the immune system. Bifidobacterial species-specific effects on dendritic cell activation were observed in that CD83 expression was increased and interleukin-10 production was induced by *Bifidobacterium bifidum, B longum,* and *Bifidobacterium pseudocatenulatum.* At least one of these species was detected in the feces of 40 out of 46 children from New Zealand and the United Kingdom, but only in a few (*B longum*: 2 of 32) of the samples from infants living in Ghana. *B infantis,* common in the feces of babies from Ghana, failed to produce these effects. It was concluded that *B infantis* would not trigger the activation of dendritic cells, thus effectively down-regulating the immune response and favoring immunological tolerance. On the other hand, *B bifidum, B longum,* and *B pseudocatenulatum* likely would drive the development of T helper 2 cells implicated in allergic responses in susceptible infants (basis not known). Further investigations of the molecular interplay between bifidobacteria, human dendritic cells, T cells and allergens, are needed. These research efforts follow in the footsteps of Dubos and colleagues, who coined the term biological Freudianism and gave as example the results of their work in which the composition of the gut microbiota of infant mice was shown to influence susceptibility to endotoxin in adulthood: a lasting effect of early environmental influences on host physiology [92]. Work with bifidobacteria and the neonatal immune system may provide the first clear evidence of immunological Freudianism in relation to human diseases.

RESEARCH ROAD MAP

An appreciation of the principles of microbial ecology is essential to understanding the relationship between the gut microbiota and the human host. The gut microbiota of people is a bacterial community of impressive biodiversity that is individualistic, interactive, self-regulating, and homeostatic. The community is metabolically active, degrading polymeric substances derived from human secretions and food and producing predictable ratios of major metabolic end products that do not vary between people. The phylogenetic make-up of the community has been elucidated reliably using nucleic acid-based methods and is dominated by members of the *Bacteroides-Prevotella, Eubacterium rectale-Clostridium coccoides,* and *C leptum* divisions. Bacterial and community genomics and immunological approaches hold the key to deriving new knowledge about the gut community and its relationship with the human host. This relationship can be likened to a magnificent iceberg. In analogy, the tip of the iceberg, which is visible above water, represents the physiological attributes of the host. The gut microbiota, unseen and unfelt, hidden from casual observation like the bulk of the iceberg below the surface of the sea, condition and shape their host for better or worse.

The signposts on the road map of gut microbiota research of particular importance to the practice of medicine are revealed by the following questions:

- Is there a mucosa-associated microbiota, and are there differences of diagnostic or pathogenic significance in their composition in specific diseases of the gut?
- What are the substances produced by the microbiota that fuel chronic gut inflammation in IBD?
- What are the substances produced by the microbiota of the neonatal gut that are important in the development of the neonatal immune system?

The answers to these questions provided by gut microbiota investigations likely will provide the insight from which new diagnostic, therapeutic, and even prophylactic solutions to diseases, initiated or fueled by immunological phenomena in the gut, will be obtained.

References

[1] Luckey TD. Germ-free life and gnotobiology. New York: Academic Press; 1963.
[2] Rettger LF, Levy MN, Weinstein L, et al. *Lactobacillus acidophilus* and its therapeutic application. New Haven (CT): Yale University Press; 1935.
[3] Savage DC. Microbial biota of the human intestine: a tribute to some pioneering scientists. Curr Issues Intest Microbiol 2001;2:1–15.
[4] Hungate RE. The rumen and its microbes. New York: Academic Press; 1966.
[5] Holdeman LV, Moore WC. Roll-tube technique for anaerobic bacteria. Am J Clin Nutr 1972;25:1314–7.
[6] Finegold SM. Anaerobic bacteria in human disease. New York: Academic Press; 1977.
[7] Aranki A, Freter R. Use of anaerobic glove boxes for the cultivation of strictly anaerobic bacteria. Am J Clin Nutr 1972;25:1329–34.
[8] Summanen P, Baron EJ, Citron DM, et al. Wadsworth anaerobic bacteriology manual. Belmont (CA): Star Publishing Company; 1993.
[9] Finegold SM, Sutter VL. Fecal flora in different populations, with special reference to diet. Am J Clin Nutr 1978;31:S116–22.
[10] Moore WEC, Cato EP, Holdeman LV. Some current concepts in intestinal bacteriology. Am J Clin Nutr 1978;31:S33–42.
[11] Moore WEC, Holdeman LV. Special problems associated with the isolation and identification of intestinal bacteria in fecal flora studies. Am J Clin Nutr 1974;27:1450–5.
[12] Drasar BS, Barrow PA. Intestinal microbiology. Washington (DC): American Society for Microbiology; 1985.
[13] O'Sullivan DJ. Methods of analysis of the intestinal microflora. In: Tannock GW, editor. Probiotics: a critical review. Wymondham (UK): Horizon Scientific Press; 1999. p. 23–44.
[14] Tannock GW, Munro K, Harmsen HJM, et al. Analysis of the fecal microflora of human subjects consuming a probiotic containing *Lactobacillus rhamnosus* DR20. Appl Environ Microbiol 2000;66:2578–88.
[15] Sait M, Hugenholtz P, Janssen PH. Cultivation of globally distributed soil bacteria from phylogenetic lineages previously only detected in cultivation-independent surveys. Environ Microbiol 2002;4:654–66.
[16] Woese CR. Bacterial evolution. Microbiol Rev 1987;51:221–71.

[17] Suau A, Bonnet R, Sutren M, et al. Direct analysis of genes encoding 16S rRNA from complex communities reveals many novel molecular species within the human gut. Appl Environ Microbiol 1999;65:4799–807.
[18] Franks AH, Harmsen HJM, Raangs GC, et al. Variations of bacterial populations in human feces measured by fluorescent in situ hybridisation with group-specific 16S rRNA-targeted oligonucleotide probes. Appl Environ Microbiol 1998;64:3336–45.
[19] Sghir A, Gramet G, Suau A, et al. Quantification of bacterial groups within the human fecal flora by oligonucleotide probe hybridisation. Appl Environ Microbiol 2000;66:2263–6.
[20] Zoetendal E, Akkermans AD, DeVos WM. Temperature gradient gel electrophoresis analysis of 16S rRNA from human fecal samples reveals stable and host-specific communities of active bacteria. Appl Environ Microbiol 1998;64:3854–9.
[21] Kaplan CW, Astaire JC, Sanders ME, et al. 16S ribosomal DNA terminal restriction fragment pattern analysis of bacterial communities in feces of rats fed *Lactobacillus acidophilus* NCFM. Appl Environ Microbiol 2001;67:1935–9.
[22] Wilson KH, Blitchington RB. Human colonic biota studied by ribosomal DNA sequence analysis. Appl Environ Microbiol 1996;62:2273–8.
[23] Requena T, Burton J, Matsuki T, et al. Identification, detection and enumeration of human *Bifidobacterium* species by PCR targeting the transaldolase gene. Appl Environ Microbiol 2002;68:2420–7.
[24] Stebbings S, Munro K, Simon MA, et al. Comparison of the faecal microflora of patients with ankylosing spondylitis and controls using molecular methods of analysis. Rheumatology 2002;41:1395–401.
[25] Harmsen JM, Welling GW. Fluorescence in situ hybridisation as a tool in intestinal bacteriology. In: Tannock GW, editor. Probiotics and prebiotics: where are we going? Wymondham (UK): Caister Academic Press; 2002. p. 41–58.
[26] Zoetendal EG, Akkermans AD, Akkermans-van Vliet WM, et al. The host genotype affects the bacterial community in the human gastrointestinal tract. Microb Ecol Health Dis 2001;13:129–34.
[27] Tannock GW. The intestinal microflora. In: Fuller R, Perdigon G, editors. Gut flora, nutrition, immunity, and health. Oxford (UK): Blackwell Publishing Limited; 2003. p. 1–23.
[28] Tannock GW, Munro K, Bibiloni R, et al. Impact of the consumption of oligosaccharide–containing biscuits on the fecal microbiota of humans. Appl Environ Microbiol 2004;70:2129–36.
[29] Felske A, Rheims H, Wolerink A, et al. Ribosome analysis reveals prominent activity of an uncultured member of the class *Actinobacteria* in grassland soils. Microbiology 1997;143:2983–9.
[30] Cummings JH, Macfarlane GT. The control and consequences of bacterial fermentation in the human colon. J Appl Bacteriol 1991;70:443–59.
[31] Podolsky DK. Inflammatory bowel disease. N Engl J Med 2002;347:417–29.
[32] Sartor RB. The influence of the normal microbial flora on the development of chronic mucosal inflammation. Res Immunol 1997;148:567–76.
[33] Sartor RB, Rath HC, Lichtman SN, et al. Animal models of intestinal and joint inflammation. Baillieres Clin Rheumatol 1996;10:55–76.
[34] Duchmann R, Kaiser I, Hermann E, et al. Tolerance exists towards resident intestinal flora but is broken in active inflammatory bowel disease (IBD). Clin Exp Immunol 1995;102: 448–55.
[35] Tannock GW. The lactic microflora of pigs, mice and rats. In: Wood BJB, editor. The lactic acid bacteria. Volume 1. The lactic acid bacteria in health and disease. London: Elsevier Applied Science; 1992. p. 21–48.
[36] Savage DC, Dubos R, Schaedler RW. The gastrointestinal epithelium and its autochthonous bacterial flora. J Exp Med 1968;127:67–76.
[37] Tannock GW. Demonstration of mucosa-associated microbial populations in the colons of mice. Appl Environ Microbiol 1987;53:1965–8.

[38] Swidsinski A, Ladhoff A, Pernthaler A, et al. Mucosal flora in inflammatory bowel disease. Gastroenterol 2002;122:44–54.
[39] Ott SJ, Musfeldt M, Wenderoth DF, et al. Reduction in diversity of the colonic mucosa associated bacterial microflora in patients with active inflammatory bowel disease. Gut 2004;53:685–93.
[40] Dal Bello F, Walter J, Hammes WP, et al. Increased complexity of the species composition of lactic acid bacteria in human feces revealed by alternative incubation condition. Microb Ecol 2003;45:455–63.
[41] Walter J, Hertel C, Tannock GW, et al. Detection of *Lactobacillus, Pediococcus, Leuconostoc,* and *Weissella* species in human feces by using group-specific PCR primers and denaturing gradient gel electrophoresis. Appl Environ Microbiol 2001;67: 2578–85.
[42] Dubos R, Schaedler RW, Costello R, et al. Indigenous, normal, and autochthonous flora of the gastrointestinal tract. J Exp Med 1965;122:67–76.
[43] Savage DC. Associations and physiological interactions of indigenous microorganisms and gastrointestinal epithelia. Am J Clin Nutr 1972;25:1372–9.
[44] Tannock GW. A special fondness for lactobacilli. Appl Environ Microbiol 2004;70: 3189–94.
[45] Guan LL, Hagen KE, Tannock GW, et al. Detection and identification of *Lactobacillus* species in crops of broilers of different ages by using PCR-denaturing gradient gel electrophoresis and amplified ribosomal DNA restriction analysis. Appl Environ Microbiol 2003;69:6750–7.
[46] Knarreborg A, Simon MA, Engberg RM, et al. Effects of dietary fat source and subtherapeutic levels of antibiotic on the bacterial community in the ileum of broiler chickens at various ages. Appl Environ Microbiol 2002;68:5918–24.
[47] Fuller R, Brooker BE. Lactobacilli which attach to the crop epithelium of the fowl. Am J Clin Nutr 1974;27:1305–12.
[48] Walter J, Heng NCK, Hammes WP, et al. Identification of *Lactobacillus reuteri* genes specifically induced in the mouse gastrointestinal tract. Appl Environ Microbiol 2003; 69:2044–51.
[49] Walter J, Chagnaud P, Tannock GW, et al. A high-molecular-mass surface protein (Lsp) and methionine sulfoxide reductase B (MsrB) contribute to the ecological performance of *Lactobacillus reuteri* in the gut of mice. Appl Environ Microbiol 2005;71:979–86.
[50] Tannock GW. Probiotics: time for a dose of realism. Curr Issues Intest Microbiol 2003;4: 33–42.
[51] Alander M, Satokari R, Korpela R, et al. Persistence of colonization of human colonic mucosa by a probiotic strain, *Lactobacillus rhamnosus* GG, after oral consumption. Appl Environ Microbiol 1999;65:351–4.
[52] Dunne C, Murphy L, Flynn S, et al. Probiotics: from myth to reality. Demonstration of functionality in animal models of disease and in human clinical trials. Antonie van Leewenhoek 1999;76:279–92.
[53] Satokari RM, Vaughan EE, Akkermans ADL, et al. Bifidobacterial diversity in human feces detected by genus-specific PCR and denaturing gradient gel electrophoresis. Appl Environ Microbiol 2001;67:504–13.
[54] Spanhaak S, Havenaar R, Schaafsma G. The effect of consumption of milk fermented by *Lactobacillus casei* strain Shirota on the intestinal microflora and immune parameters in humans. Eur J Clin Nutr 1998;52:899–907.
[55] Saavedra JM. Probiotics and infectious diarrhea. Am J Gastroenterol 2000;95:S16–8.
[56] Alexander M. Microbial ecology. New York: John Wiley and Sons; 1971.
[57] Hardin G. The competitive exclusion principle. Science 1960;131:1292–7.
[58] Kassen R, Rainey PB. The ecology and genetics of microbial diversity. Annu Rev Microbiol 2004;58:207–31.

[59] Kimura K, McCartney AL, McConnell MA, Tannock GW. Analysis of fecal populations of bifidobacteria and lactobacilli and investigations of the immunological responses of their human hosts to the predominant strains. Appl Environ Microbiol 1997;63:3394–8.
[60] McCartney AL, Wang W, Tannock GW. Molecular analysis of the composition of the bifidobacterial and lactobacillus microflora of humans. Appl Environ Microbiol 1996;62:4608–13.
[61] Crittenden RG. Prebiotics. In: Tannock GW, editor. Probiotics: a critical review. Wymondham (UK): Horizon Scientific Press; 1999. p. 141–56.
[62] Gibson GR, Beatty ER, Wang X, et al. Selective stimulation of bifidobacteria in the human colon by oligofructose and inulin. Gastroenterology 1995;108:975–82.
[63] Bouhnik Y, Flourie B, Riottot M, et al. Effects of fructo–oligosaccharides ingestion on fecal bifidobacteria and selected metabolic indexes of colon carcinogenesis in healthy humans. Nutr Cancer 1996;26:21–9.
[64] Gibson GR, Beatty ER, Wang X, et al. Selective stimulation of bifidobacteria in the human colon by oligofructose and inulin. Gastroenterol 1995;108:975–82.
[65] Harmsen HJM, Raangs GC, Franks AH, et al. The effect of the prebiotic inulin and the probiotic *Bifidobacterium longum* on the fecal microflora of healthy volunteers measured by FISH and DGGE. Microb Ecol Health Dis 2002;14:211–9.
[66] Kruse HP, Kleesen B, Blaut M. Effects of inulin on faecal bifidobacteria in human subjects. Br J Nutr 1999;82:375–82.
[67] Rao AV. Dose-response effects of inulin and oligofructose on intestinal bifidogenesis effects. J Nutr 1999;129:1442S–5S.
[68] Tuohy KM, Kolida S, Lustenberger AM, et al. The prebiotic effects of biscuits containing partially hydrolysed guar gum and fructo–oligosaccharides—a human volunteer study. Br J Nutr 2001;86:341–8.
[69] Schell MA, Karmirantzou M, Snel B, et al. The genome sequence of *Bifidobacterium longum* reflects its adaptation to the human gastrointestinal tract. Proc Natl Acad Sci U S A 2002;99:14422–7.
[70] Krinos CM, Coyne MJ, Weinacht KG, et al. Extensive surface diversity of a commensal microorganism by multiple DNA inversions. Nature 2001;414:555–8.
[71] Xu J, Bjursell MK, Himrod J, et al. A genomic view of the human-*Bacteroides thetaiotaomicron* symbiosis. Science 2003;299:2074–6.
[72] Hopkins MJ, Englyst HN, Macfarlane S, et al. Degradation of cross-linked and non-cross–linked arabinoxylans by the intestinal microbiota in children. Appl Environ Microbiol 2003;69:6354–60.
[73] Roberton AM, Corfield AP. Mucin degradation and its significance in inflammatory conditions of the gastrointestinal tract. In: Tannock GW, editor. Medical importance of the normal microflora. Dordrecht, The Netherlands: Kluwer Academic Publishers; 1999. p. 222–61.
[74] Rondon MR, August PR, Bettermann AD, et al. Cloning the soil metagenome: a strategy for accessing the genetic and functional diversity of uncultured microorganisms. Appl Environ Microbiol 2000;66:2541–7.
[75] Schloss PD, Handelsman J. Biotechnological prospects from metagenomics. Curr Opin Biotechnol 2003;14:303–10.
[76] Walter J, Mangold M, Tannock GW. Construction, analysis and β-glucanase screen of a bacterial artificial chromosome library from the large bowel microbiota of mice. Appl Environ Microbiol 2005;71:2347–54.
[77] Xu J, Gordon JI. Honor thy symbionts. Proc Natl Acad Sci U S A 2003;100:10452–9.
[78] Bealmear PM, South MA, Wilson R. David's story: the gift of 12 years, 5 months, and 1 day. In: Germ-free research: microflora control and its application to the biomedical sciences. New York: Alan R Liss Incorporated; 1985. p. 475–89.

[79] Gordon HA, Pesti L. The gnotobiotic animal as a tool in the study of host–microbial relationships. Bacteriol Rev 1971;35:390–429.
[80] Hooper LV, Wong MH, Thelin A, et al. Molecular analysis of commensal host–microbial relationships in the intestine. Science 2001;291:881–4.
[81] Hooper LV, Xu J, Falk PG, et al. A molecular sensor that allows a gut commensal to control its nutrient foundation in a competitive ecosystem. Proc Natl Acad Sci U S A 1999; 96:9833–8.
[82] Stappenbeck TS, Hooper LV, Gordon JI. Developmental regulation of intestinal angiogenesis by indigenous microbes via Paneth cells. Proc Natl Acad Sci U S A 2002;99:15451–5.
[83] Backhed F, Ding H, Wang T, et al. The gut microbiota as an environmental factor that regulates fat storage. Proc Natl Acad Sci U S A 2004;101:15718–23.
[84] Tannock GW. Normal microflora. An introduction to microbes inhabiting the human body. London: Chapman and Hall; 1995.
[85] Tissier H. Repartition des microbes dans l'intestin du nourrisson. Ann Inst Pasteur (Paris) 1905;19:109–23.
[86] Harmsen HJM, Wildeboer AC, Raangs GC, et al. Analysis of intestinal flora development in breast-fed and formula-fed infants by using molecular identification and detection methods. J Pediatr Gastroenterol Nutr 2000;30:61–7.
[87] Hopkin JM. Mechanisms of enhanced prevalence of asthma and atopy in developed countries. Curr Opin Immunol 1997;9:788–92.
[88] Murray CS, Woodcock A. Gut microflora and atopic disease. In: Tannock GW, editor. Probiotics and prebiotics: where are we going? Wymondham (UK): Caister Academic Press; 2002. p. 239–61.
[89] Sepp E, Julge K, Vasar M, et al. Intestinal microflora of Estonian and Swedish infants. Acta Paediatr 1997;86:956–61.
[90] Ouwehand AC, Isolauri E, He F, et al. Differences in *Bifidobacterium* flora composition in allergic and healthy infants. J Allergy Clin Immunol 2001;108:144–5.
[91] Young SL, Simon MA, Baird MA, et al. Bifidobacterial species differentially affect expression of cell surface markers and cytokines of dendritic cells harvested from cord blood. Clin Diag Lab Immunol 2004;11:686–90.
[92] Schaedler RW, Dubos RJ. The fecal flora of various strains of mice. Its bearing on their susceptibility to endotoxin. J Exp Med 1962;115:1149–60.

Gastroenterol Clin N Am 34 (2005) 383–399

GASTROENTEROLOGY CLINICS
OF NORTH AMERICA

Physiologic and Pathologic Interactions of Bacteria with Gastrointestinal Epithelium

Marko A. Kalliomäki, MD, PhD[a], W. Allan Walker, MD[b,*]

[a]Department of Pediatrics, Turku University Hospital and University of Turku, Kiinamyllykatu 4-8, FIN-20521, Turku, Finland

[b]Mucosal Immunology Laboratory, Massachusetts General Hospital for Children and Harvard Medical School, 114 16th Street, Charlestown, MA 02129, USA

A single layer of epithelial cells lining the human gastrointestinal (GI) tract separates the host's internal from external milieu. Recent scientific data suggest that the gut epithelium serves not only as a physical barrier and an organ system for digestion and absorption of life-sustaining nutrients but also as the front-line conductor for the immune system. Physiologic interactions between gut epithelium and its abundant indigenous bacteria create an important basis for intestinal homeostasis. Disruption of that delicate equilibrium can result in intestinal inflammation (eg, inflammatory bowel disease, gastroenteritis, or pathogen translocation, leading to sepsis). In addition, bacterial pathogens have developed several mechanisms to manipulate the enterocyte's functions for their own survival.

Because this issue of the *Gastrointestinal Clinics of North America* is devoted to probiotics, prebiotics and commensals, that is nonpathological intestinal bacteria, this article concentrates on the most recent findings regarding their role in the gut, especially the physiologic interaction between these bacteria and underlying GI epithelium. Because of space limitations, this article does not consider bacterial interaction with the gastric epithelium, although their potential importance in modulating the interaction between the gastric epithelium and *Helicobacter pylori* must be stressed. Although some of most recent findings on pathological interactions in the gut are highlighted, several comprehensive reviews on this subject are cited for the reader who would prefer to obtain a more comprehensive understanding of this topic [1–9].

This article was supported by funding from the National Institutes of Health (R37 HD12437, RO1 DK70260, P30 DK40561 and PO1 DK33506, WAW). The Mucosal Immunology Laboratory has unrestricted funding from Wyeth Ayerst Nutritionals International.

*Corresponding author. *E-mail address:* wwalker@partners.org (W.A. Walker).

0889-8553/05/$ – see front matter

doi:10.1016/j.gtc.2005.05.007
gastro.theclinics.com

THE COMPOSITION OF THE GUT'S EPITHELIAL SURFACE

An absorptive mucosa is the typical surface of the small intestine. To maximize absorptive surface area, the small intestine has evolved several adaptations, including the mucosa folded into finger-like projections called villi with intervening short glands called crypts of Lieberkühn. Thousands of microvilli per single enterocyte further increase surface area for absorption across the plasma membrane exposed to the intestinal lumen. The epithelial lining of the large intestine also contains microvilli [10]. Consequently, the total surface area of the GI tract is huge, exceeding that of the skin by approximately 100-fold. Besides its extensive surface area, an active renewal of epithelium is a hallmark of the GI epithelium. The epithelium is replaced every 3 to 5 days. The renewal process involves stem cell proliferation and cell migration. In the small and large intestine, cells differentiate into enterocytes, goblet cells, or enteroendocrine cells during their migration from crypts upward [10,11]. This enormous surface area and active cell turnover create a challenge in protecting the gut surface from its hostile environment.

Enterocytes and goblet cells are the major cell types of the intestinal villi and colonic epithelium. Beyond absorptive function, these enterocytes participate in innate and adaptive immune protection. They secrete plasma cell-derived polymeric immunoglobulin A (pIgA), the largest noninflammatory defense system in people, and β-defensins, small cationic peptides, which, besides their antimicrobial activities, are chemotactic for dendritic and T cells [4,11–14]. Intestinal goblet cells produce and secrete mucous, which provides a mucopolysaccharide barrier between the epithelium and the contents of the lumen. This barrier also helps prevent microbial attachment and colonization [11]. The small intestine also contains Paneth's cells, which are specialized epithelial cells found in clusters at the base of the crypts of Lieberkühn [14]. These cells are packed with eosinophilic granules that contain several antimicrobial peptides, including lysozyme, secretory phospholipase A_2, α-defensins, trypsin, and angiogenins. The close proximity of Paneth's cells to epithelial stem cells suggests that they function to protect stem cells from luminal pathogens by releasing their antimicrobial peptides [4,14] on contact with the organism. Unlike other epithelial lineages of the small intestine, Paneth's cells undergo terminal differentiation during downward migration from the stem cell progenitor toward the base of the crypt [11]. Their lifespan is also much longer than that of other intestinal epithelial cells (20 to 25 days).

Follicle-associated epithelium (FAE), overlying the uppermost part of the Peyer's patches, is a specialized intestinal epithelium without mucus, crypts or microvillus, or organelles. These epithelia do not secrete IgA. Literally, the FAE overlies the gut-associated lymphatic tissue (GALT), containing large numbers of B cells, T cells, macrophages, and dendritic cells [15]. The most remarkable feature of the FAE is the presence of specialized antigen-sampling membranous or microfold (M) cells. These cells function primarily to sample dietary and microbial antigens for the underlining GALT [16]. Recent studies have shown that the M cell-mediated antigen transport across the intestinal mucosal barrier can be up-regulated by interaction with bacteria without altering the barrier integrity [17].

BACTERIAL SENSING IN THE GUT EPITHELIUM

The GI mucosa constantly encounters large quantities of microbes, which are, by and large, nonpathogenic. Indeed, it has been estimated that the number of the bacterial inhabitants of the GI tract exceeds that of total body eukaryotic cells by a factor of 10 [18]. To sense that myriad of microorganisms, the GI epithelium expresses various pattern recognition receptors (PRRs) including several forms of toll-like receptors (TLRs) [19] that recognize specific conserved pathogen-associated molecular patterns (PAMPs) on microorganisms (see Appendix 1 for further information on terms and abbreviations used in this article). TLR2, TLR3, TLR4, TLR5, TLR6, and TLR9 have been identified in the gut epithelium by earlier studies [20–22]. Recently, messenger RNA coding for TLRs 1–6, 8 and 9 was detected by using a transcription-polymerase chain reaction (RT-PCR) technique in freshly isolated human primary colonic epithelial cells [23]. The expression pattern closely resembled that of the intestinal epithelial cell lines Colo205 and SW480 [23]. None of the TLR microbial pattern ligands are unique for pathogens, because they also are expressed by nonpathogenic organisms. Thus, they can be referred to as commensal-associated molecular patterns (CAMPs) [24]. This observation has important implications as one considers the role of physiologic microbial epithelial cross-talk in the context of gut protection. Structurally, all human TLRs are type I transmembrane receptors with leucine-rich extracellular repeats and an intracellular signaling domain known as the Toll-IL-2 receptor (TIR) domain [25]. Typically, the intracellular TIR domain of TLRs initiate signaling cascades that result in production of inflammatory cytokines by activating transcription factors nuclear factor κB (NF-κB), activator protein 1 (AP-1), and interferon regulatory factor (IRF) 3 [25]. A cartoon presentation of different TLRs, their ligands, and activation in intestinal epithelial cells is depicted in Fig. 1.

Intestinal epithelial cells express nucleotide oligomerization domain 1 (NOD1) and nucleotide oligomerization domain 2 (NOD2) pattern recognition receptors in their cytosol, recently renamed as caspase activating recruitment domain (CARD) 4 and 15, respectively. The former recognizes the diaminopimelic acid component of peptidoglycan in gram-negative bacteria, whereas the latter is a pattern recognition receptor for muramyl dipeptide, a breakdown product of the peptidoglycan component of all bacterial cell walls. Following microbial ligand recognition by NOD1/CARD4 and NOD2/CARD15, proinflammatory genes are activated by means of NF-κB [26–29] (see Fig. 1). Susceptibility to Crohn's disease, especially to terminal ileal involvement, has been linked to mutations in NOD2/CARD15 gene [30–34]. NOD2/CARD15 has been shown to be expressed in Paneth's cells of normal duodenum, jejunum, and terminal ileum; the greatest staining intensity was found in terminal ileal crypts. In patients who have Crohn's disease, the expression of NOD2/CARD15 was increased in Paneth's cells within the

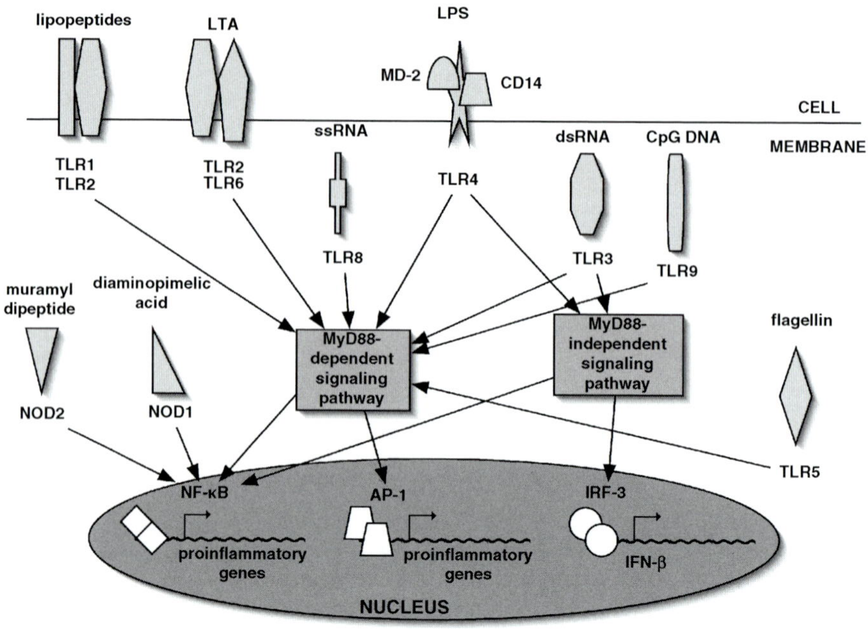

Fig. 1. A simplified presentation of pattern recognition receptors, their typical ligands, and intracellular activation in the intestinal epithelial cell. Toll-like receptor (TLR) 1, TLR2, TLR5 (basolateral), and TLR6 are located on the cell membrane, whereas TLR3, TLR8, TLR9, nucleotide oligomerization domain (NOD) 1, and NOD2 are located in the cytosol. TLR4 is shown to be located on the apical cell membrane, but some mouse studies suggest TLR4 also may reside in the Golgi apparatus of the intestinal epithelial cell, at least in the small intestine. MD-2 and CD14 are essential molecules required for TLR4 activation. Only one ligand per receptor is mentioned. although many TLRs can sense several different ligands. TLR3 and TLR4 can activate both myeloid differentiation primary-response protein 88 (MyD88)-dependent and -independent signaling pathways, whereas other TLRs use only the former pathway. The MyD88-dependent signaling pathway activates expression of proinflammatory genes by means of nuclear factor κB (NF-κB) and activator protein 1 (AP-1), whereas the MyD88-independent signaling results in expression of interferon beta (IFN-β) by means of IFN regulatory factor 3 (IRF-3) and that of proinflammatory genes by means of NF-κB. *Abbreviations:* CpG DNA, unmethylated cytosine-guanine rich DNA; dsRNA, double-stranded RNA; LPS, lipopolysaccharide; LTA, lipoteichoic acid; ssRNA, single-stranded RNA.

terminal ileum. The expression also was noted in metaplastic Paneth's cells within the colon [35,36]. Recently, patients who have Crohn's disease with NOD2/CARD15 mutations have been shown to have decreased mucosal α-defensin expression by real-time quantitative RT-PCR in affected ileum and colon as compared with healthy controls or patients who have Crohn's disease with wild-type NOD2/CARD15 [37,38]. The authors of that study recently published their general hypothesis for the pathogenesis of Crohn's disease as a defensin deficiency syndrome [39]. According to this hypothesis, a diminished α-defensin expression in Paneth's cells and β-defensin expression in enterocytes result in an impaired intestinal barrier to intestinal microbes, leading to development of ileal and colonic Crohn's disease [39]. Regulation of human β-defensin-2 (hBD-2) expression in intestinal epithelial cells has been shown to be dependent on TLR signaling [40]. TLR4 and MD-2 (*v-myb* regulated gene 2) are required for LPS-mediated expression of hBD-2 in intestinal epithelial cells, whereas TLR2 and TLR6 are prerequisites for peptidoglycan-induced expression of hBD-2 in these same cells. Downstream, both NF-κB and AP-1 transactivation are needed for optimal expression of this enterocyte defensin [40].

Continuous activation of innate immune responses with the production of inflammatory cytokines that would stimulate the adaptive immune system could be devastating to the host. Therefore, several characteristics of the gut epithelium have been suggested that are thought to prevent an inappropriate and dangerous activation of inflammatory cascades and to augment development of tolerance toward indigenous microbiota. First, PRRs are expressed either in the cytoplasm (TLR3, TLR8, TLR9, NOD1/CARD4, NOD2/CARD15) or basolaterally (TLR5). That is, bacteria or their molecular patterns have to invade across the epithelium to encounter the PRRs [21,25,41]. Second, some studies have found only limited expression of TLR2, TLR4, MD-2, the essential accessory protein of TLR4, and CD14, the coreceptor molecule of the TLR4, in human colon carcinoma cell lines, implying that immune responses toward their microbial ligands, bacterial lipoteichoic acid (LTA), and lipopolysaccharide (LPS), respectively, can be diminished [22,42,43]. Conflicting findings have been reported, however [20,44]. Discrepant findings among in vitro studies likely are explained best by the diverse function of various human colon carcinoma cell lines under different experimental conditions [44]. Studies with the mouse small intestinal epithelial cell line m-IC_{c12} have shown that TLR4 resides in conjunction with internalized LPS in the Golgi apparatus of intestinal crypt epithelial cells [45]. These same cells also express MD-2 and CD-14. Moreover, the TLR4 adaptor protein myeloid differentiation factor 88 (MyD88) necessary for signaling, was recruited to the Golgi apparatus after LPS stimulation, suggesting an intracellular event [46]. Whether human intestinal epithelial cells share similar intracellular reactions needs to be determined. Several inhibitors of TLR signal transduction have been detected in enterocytes as reviewed by Haller and Jobin [1]. These include the regulatory cytokines interleukin (IL)-10 and transforming growth factor-β and intrinsic regulators such as toll-interacting protein (Tollip), single immunoglobulin IL-1R-related protein (SIGIRR), suppressor of cytokine signaling (SOCS), peroxisome proliferator activated receptor-γ (PPAR-γ), the zinc finger protein A20, and IL-1 receptor-associated kinase-M (IRAK-M) [1].

A crucial role of Tollip in the regulation of innate immune responses of the intestinal epithelium has been reported [23]. Otte and colleagues studied the effects of exposure of LTA and LPS on expression and function of TLRs in colonic epithelial cell lines (Colo205, SW480, T84). They showed that short-term stimulation with either ligand activated proinflammatory signaling cascaded and increased IL-8 secretion, whereas prolonged exposure (24 hours) resulted in a state of hyporesponsiveness (tolerance) [23]. In addition, challenge with either substance did not reactivate the cells (cross-tolerance). In hyporesponsive cells, total expression of TLRs was unaltered, although TLR surface expression was decreased. Tollip expression was increased in these cells, and up-regulation of the Tollip in the intestinal epithelial cells resulted in a decreased proinflammatory response [23]. On a condition that these mechanisms are operative in vivo, these findings imply that until the state of hyporesponsiveness in the gut epithelium is established, encounters of intestinal microbes with GI epithelial cells may result in significant inflammatory responses. In neonates, that kind of responses might be regulated further by breastfeeding, because breastmilk contains functional soluble TLR2 and CD14 [47,48]. Developmental regulation of inhibitory κB (IκB) genes, key regulators of NF-κB-dependent inflammatory pathways [25], has been reported in intestinal epithelium [49]. In that study, the flagellin-induced IL-8 response of an immature human enterocyte cell line (H4 cells) to bacterial infection was significantly higher than that of a mature enterocyte cell line (T84), regardless of a comparable expression of TLR5 in these cells. Transfection studies confirmed that the decreased level of the inhibitory protein IκB in immature H4 cells resulted in the enhanced proinflammatory response [49]. This characteristic of the immature human intestinal epithelial cell may contribute to the pathogenesis of age-specific inflammatory bowel diseases such as necrotizing enterocolitis [49]. Expression of negative regulators of TLR signal transduction in immature enterocytes remains to be elucidated.

COMMENSAL MICROBIAL–ENTEROCYTE CROSS-TALK: A KEY TO INTESTINAL HOMEOSTASIS

The use of an in vivo gnotobiotic mouse model for the study of the intestinal microbial ecosystem has provided insight into the physiologic consequences of host–microbe interactions in the gut. In this model, germ-free mice were monocolonized with a bacterial strain common to the mouse and human distal intestine, *Bacteroides thetaiotaomicron* [50–52]. Ileal transcriptional responses to colonization were measured by cDNA microarrays and confirmatory real-time quantitative RT-PCRs. Laser-capture microdissection was used to study intestinal cellular origins of selected responses. The results showed that *B thetaiotaomicron* regulated production of ileal epithelial fucosylated glycans for its own nutritional benefit and survival, that is, to promote its own adherence and colonization in the epithelium [50]. On the host side, the mono-association of germ-free mice with *B thetaiotaomicron* resulted in a modulated expression of genes regulating a range of intestinal functions, including nutrient absorption, angiogenesis, metabolism of endogenous toxins, and fortification of the innate immune system [51,53,54]. The colonization induced the expression of a previously uncharacterized mouse angiogenin, Ang4, which was produced by Paneth's cells and has bactericidal activity against intestinal microbes [54]. At least some of the colonization-associated changes in gene expression were different when strains other than *B thetaiotaomicron* were used, implying that the host's physiology may be affected by differences in the composition of indigenous microflora [51].

A recent study provides further evidence that the interaction between commensals and the host has an indispensable role in maintaining intestinal epithelial homeostasis. In this study, intestinal inflammation was induced by dextran sulfate sodium (DSS) in wild-type mice, in mice deficient in MyD88 (MyD88$^{-/-}$), and in mice deficient in TLR2 (TLR2$^{-/-}$) and TLR4 (TLR4$^{-/-}$) [55]. MyD88$^{-/-}$ knockouts disrupt signal transduction that is used by all cell toll-like receptors. More severe epithelial injuries and ulcerations were found in the colon of MyD88$^{-/-}$ mice compared with wild-type mice. MyD88-deficient animals also showed greater mortality and morbidity upon administration of DSS, although all wild-type animals survived. A histopathological scoring of the colon of these animals, their death rate, and weight loss were greater among TLR2$^{-/-}$ and TLR4$^{-/-}$ mice compared with wild-type mice, but less than among MyD88$^{-/-}$ mice [55]. Complete depletion of intestinal commensals achieved by a 4-week regimen of vancomycin, neomycin, metronidazole, and ampicillin in wild-type mice before administration of DSS resulted in similar severe colon bleeding, morbidity, and mortality as seen in MyD88-deficient mice. These consequences could be prevented in the commensal-depleted animals if they were supplemented orally with either LTA or LPS [55]. These authors suggested that commensal bacterial interaction with enterocyte or lamina propria lymphocyte by means of toll-like receptors was required for protection from these insults [55].

ENHANCEMENT OF INTESTINAL EPITHELIAL BARRIER FUNCTION BY COMMENSALS

Relative impermeability of the intestinal epithelial barrier to microorganisms represents one basis for the innate immune system in the gut. Consequently, translocation of potentially dangerous luminal antigens and detrimental effects of subsequent activation of systemic inflammatory responses can be diminished. Some probiotics, live microbial food ingredients that beneficially affect host health [56], have been shown to strengthen the intestinal epithelial barrier in vivo. *Lactobacillus rhamnosus* GG reverses the increased intestinal permeability induced by reaction of cow milk protein in suckling rats [57]. The combination of *L rhamnosus* 19070-2 and *L reuteri* DSM 12,246 was efficacious in stabilizing the intestinal mucosal barrier and reducing GI symptoms in children with atopic dermatitis [58]. There is also a positive correlation between permeability of the intestinal barrier and the severity of the atopic dermatitis [58]. A similar combination of probiotics has been shown to reduce the extent of atopic dermatitis [59]. Treatment of colitis in IL-10 gene-deficient mice with a combination of probiotics (a mixture of *Bifidobacterium longum, Bifidobacterium infantis, Bifidobacterium breve, L acidophilus, L casei, L delbrueckii* subspecies *L bulgaricus, L plantarium, Streptococcus salivarius* subspecies *Thermophilus* called

VSL#3) resulted in normalization of the colonic barrier integrity, improvement in histologic disease activity, and a reduction in mucosal secretion of proinflammatory cytokines [60]. In vitro studies have shown that epithelial barrier function and resistance to *Salmonella* invasion could be increased by exposure to a proteinaceous soluble factor secreted by the VSL#3 mixture [60,61]. VSL#3 also has been noted to significantly reduce *Salmonella*-induced alterations in the cellular cytoskeleton of the intestinal epithelium, including the distribution of the tight junction protein zonula occludens 1 (ZO-1). The effect, at least in part, was mediated by the mitogen-activated protein kinase (MAPK) signaling cascade [61]. In addition, it has been shown that the synthetic bacterial lipopeptide $Pam_3CysSK4$, a peptidoglycan, can induce activation of two protein kinase C (PKC) isoforms in the human colonic epithelial cell lines HT-29 and Caco-2 [62]. TLR2 expression was found to be essential for activating PKC by $Pam_3CysSK4$. Moreover, the activation correlated with apical tightening and sealing of tight junction-associated ZO-1, implying that the interaction of TLR2 and peptidoglycan increases ZO-1-associated intestinal epithelial barrier integrity by means of PKC signaling [62].

Commensals also have been shown to induce T cell-independent polymeric IgA responses by stimulating activation of intestinal dendritic cells that significantly enforce barrier integrity [63]. The probiotics *L rhamnosus* GG and *L plantarum* 299v, but not *L acidophilus* DDS-1, adhered effectively to HT-29 cells and increased the expression of intestinal mucins MUC2 and MUC3 [64,65]. The adherence of enteropathogenic *Escherichia coli* to this cell line was inhibited by the up-regulation of these mucins [64,65]. VSL#3 also has been demonstrated to up-regulate gene expression and production of mucins in HT-29 cells [61]. MAPK signaling pathway appears to mediate these effects [61]. The same kind of up-regulation has not seen with *E coli* Nissle 1917 stimulation, suggesting specificity of the response [61]. *E coli* Nissle 1917 and some lactobacilli, but not of the pathogenic *E coli* strains, strongly induced the expression of the antimicrobial peptide hBD-2 in Caco-2 cells in a time- and dose-dependent manner, however, suggesting another response by this probiotic [66]. Further experiments demonstrated that functional binding sites for NF-κB and AP-1 in the hBD-2 promoter region are required for induction of hBD-2 by *E coli* Nissle 1917 [66]. This novel mechanism of action of a probiotic is especially interesting in the context of the new hypothesis for the pathogenesis of Crohn's disease as a defensin deficiency syndrome [39]. Commensals and probiotics thus seem to contribute in several ways to strengthening the intestinal epithelial barrier and providing first-line host defense in the gut.

ANTI-INFLAMMATORY EFFECTS OF COMMENSALS IN INTESTINAL EPITHELIAL CELLS

Several in vitro studies have demonstrated that certain strains of commensals and probiotics elicit anti-inflammatory responses in intestinal epithelial cells.

Neish and colleagues [67] reported that a nonpathologic strain of *Salmonella* was able to attenuate IL-8 secretion elicited by pathogenic *Salmonella* in a polarized intestinal cell line T84 model. A similar inhibitory effect was noted if IL-8 activation occurred after other proinflammatory stimuli, including tumor necrosis factor α (TNF-α). This immunosuppressive effect involved inhibition of thc proinflammatory/antiapoptotic IκB/NF-κB pathway by blockage of IκB-α degradation, thus preventing disassociation and subsequent nuclear translocation of active NF-κB [67]. In a subsequent study, these authors reported that AvrA, a member of the YopJ/Avr family of secreted bacterial proteins, was the effector molecule that inhibits activation of the key inflammatory NF-κB transcription factor and contributes to apoptosis in human epithelial cells [68]. The AvrA protein was delivered to the host cell by means of a type III secretion system. These observations suggest that the mammalian host may exploit apoptosis in rapidly turning over epithelial cells as a defense mechanism [68]. Functionally analogous molecules have been characterized in plants that elicit a host defense response that results in local apoptosis of infected cells, thus limiting further infection by an invading pathogen [69].

In additional studies that support observations with nonpathogenic *Salmonella* [67], the probiotic *L rhamnosus* GG has been shown to prevent cytokine-induced apoptosis in mouse and human colon cells [70]. The effect was found to be mediated by activation of antiapoptotic Akt and protein kinase B molecules and inactivation of the proapoptotic p38 MAPK signaling cascade [70]. MAPK signal transduction pathways have been linked to a range of pathologic immune responses in the gut, and they are a novel target for anti-inflammatory therapeutic molecules [71]. Two proteins identified from *L rhamnosus* GG culture broth elicited almost identical effects as the bacterium itself. Moreover, *L acidophilus* and *L casei* also activated Akt but did not inactivate the p38 MAPK [70] pathway. Recently, *L reuteri* was reported to inhibit constitutive gene expression and production of IL-8 in T84 and HT-29 cells [72]. Similar inhibition was observed if IL-8 was induced by TNF-α or *Salmonella typhimurium.* In support of earlier *Salmonella* studies [67], this immunosuppression was mediated by blocking of the IκB/NF-κB pathway [72]. Kelly and colleagues showed that *B thetaiotaomicron* was able to significantly attenuate a proinflammatory response elicited by *S enteritidis* in intestinal epithelial cells in vitro and rat's intestinal epithelium in vivo [73]. The inhibitory effect was based on a nuclear export of RelA, a transcriptionally active NF-κB subunit, by a PPAR-γ-dependent mechanism [73]. Jijon and colleagues [73] reported that orally fed DNA from VSL#3 was able to limit colonic epithelial proinflammatory responses induced by pathogenic bacteria and improve the histologic intestinal inflammation score in IL-10-deficient mice. In vitro studies revealed that the probiotic DNA suppressed proinflammatory signaling pathways elicited by IL-8 [74]. The differing effects of commensal microorganisms and probiotics on protective/anti-inflammatory function of the intestinal epithelium are depicted in Fig. 2.

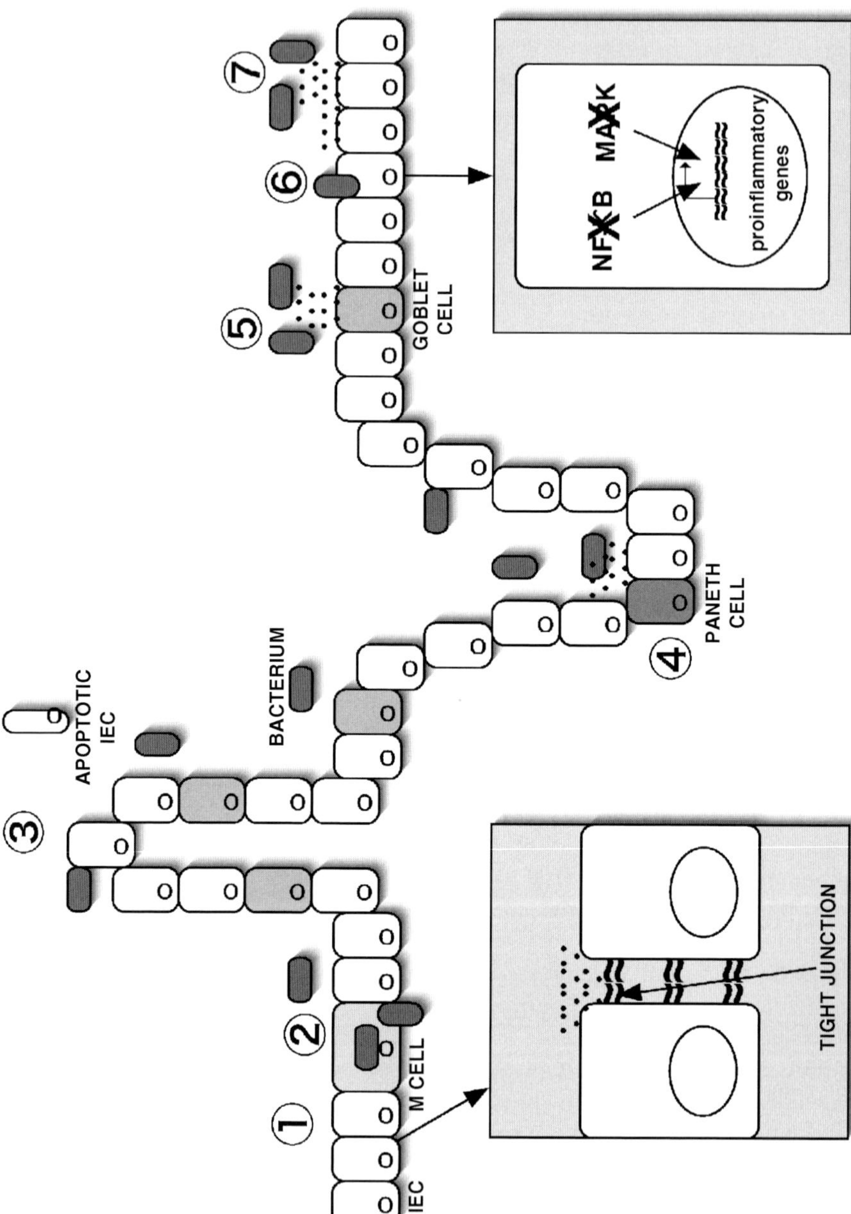
1
2
3
4
5
6
7
IEC
M CELL
APOPTOTIC IEC
BACTERIUM
PANETH CELL
GOBLET CELL
TIGHT JUNCTION
NF B
MA K
proinflammatory genes

PATHOLOGIC INTERACTIONS WITH THE GUT EPITHELIUM

In the distal small intestine and colon, virtually every ecological niche is colonized by bacteria. Successful interaction of pathogens with the GI epithelium is dependent on their ability to perturb the homeostasis of the digestive tract. Not only do commensal bacteria and probiotics interact with the gut epithelium to help establish the mucosal barrier, but pathogens also communicate with the underlying epithelium to establish their niche. Therefore, enteropathogens have developed several characteristics or methods that assist in invading and exploiting the functions of the host cell. These include secretion of bacterial enterotoxins, a type III secretion system, pathogenicity islands, and rearrangement of the host cytoskeleton [2,3,5–9].

PATHOGENICITY AND GENOMIC ISLANDS

A pathogenicity island or locus is a region of the bacterial chromosome or within a plasmid that encodes one or more virulence factors necessary for expressing human disease [75]. The difference between a commensal and a pathogen may depend entirely on the presence of such virulence factors in pathogenicity loci. Transfer of such a pathogenicity package or virulence cassette to a commensal may transform it into a pathogen. For example, disparities between indigenous and pathogenic *E coli* strains in the gut can be explained by differences in the expression of virulence factors [76]. Recent studies suggest that commensal genomes also contain pathogenicity island-type mobile genetic elements called genomic islands. Typically, these structures are comprised of several thousands of base pairs associated with transfer RNA-encoding genes [75]. These genetic islands have been thought to contribute to several processes that affect the capacity of commensals to survive in ecological niches. A transfer of this type of mobile genetic element is at least a theoretical risk that should be considered when preparing genetically modified probiotics [77].

THE TYPE III SECRETION SYSTEM

The type III secretion system (T3SS) is a virulence factor encoded by pathogenicity islands in multiple enteropathogens, including *Salmonella* species, *Shigella* species, *Yersinia* species, and *E coli* species. The gene product is a needle-like structure that bridges between bacterial and host cells, thus allowing

Fig. 2. Physiologic effects of gut microbiota on the gut epithelium. (1) Commensals have been demonstrated to strengthen the intestinal epithelial barrier integrity by modulating tight junction-associated proteins (enlargement on the left). (2) M cell-mediated antigen transport across the intestinal epithelium can be increased by interaction with bacteria. (3) Commensals can regulate apoptosis of intestinal epithelial cell (IEC). (4,5,7) Certain probiotic strains are capable of inducing expression of several different antimicrobial peptides in the gut epithelium, including Paneth's cell-derived angiogenin 4, goblet cell-derived mucins, and IEC-derived human β-defensin 2. (6) Commensals have anti-inflammatory effects on IEC's gene expression by inhibiting signaling pathways of nuclear factor κB (NF-κB) and mitogen-activated protein kinases (MAPK).

delivery of bacterial proteins (eg, effector proteins and regulators of human enterocyte gene expression) from the bacterial cytoplasm into the host cell [78]. It in essence functions as a molecular syringe. *S enterica* possess two distinct T3SSs, T3SS-1 and T3SS-2, which are encoded by *Salmonella* pathogenicity islands 1 and 2, respectively. The former is required for invasion and the latter for intracellular pathogenesis [6]. Upon secretion from the bacterial cell, some of T3SS1-derived effector proteins modify small GTPases of the host cell, triggering rearrangements in the actin cytoskeleton of the enterocyte that lead to formation of membrane ruffles and internalization of attaching *Salmonella* by a process called macropinocytosis [79]. Alternatively, *Salmonella* can traverse the GI epithelium by means of M cells [8]. The contribution of different effectors to the invasion of *S typhimurium* into epithelial cells in vitro has been evaluated [80]. Raffatellu and colleagues found that the contribution of different effectors to *S typhimurium* invasion was dependent on polarization of cells and cell types used. Invasion to nonpolarized T84 and HT-29 cells was dependent on SopE2 effector protein and SipA, SopB, and SopE2, respectively, whereas SopB, SopD, SopA, SopE2, and SipA were required for invasion into polarized T84 cells. The polarized in vitro model epithelium likely stimulates the in vivo situation more accurately [80]. Some nonpathogenic *Salmonella* strains have been shown to use T3SS to inject their inhibitory enzymatic effectors into the host cell [68,81] to reduce inflammation.

BACTERIAL ENTEROTOXINS

Noninvasive enteropathogens, such as *Vibrio cholerae*, enterotoxigenic *E coli,* enterotoxigenic *Bacteroides fragilis*, and *Clostridium difficile,* can cause diarrhea in people by delivering toxins that affect fluid homeostasis in the gut [3]. Overgrowth of *C difficile* and the release of toxin A and toxin B lead to inflammation and damage of colonic mucosa seen in antibiotic-associated diarrhea and pseudomembranous colitis [82]. These toxins inactivate the Rho guanosine triphosphatases (GTPases) RhoA, Rac1, and Cdc42 by glycosylation that leads to disruption of the actin cytoskeleton [7]. In addition, a recent cDNA microarray study evaluating the effects of toxin A in vitro revealed that the gene of the GTPase RhoB was the predominant gene up-regulated in Caco-2 and HT-29 cells, through activation of p38-MAPK pathway for pathogenicity [83]. RhoB was inactivated only partially by glycosylation, so that a significant portion of RhoB was capable of downstream signaling. These findings imply that toxin A is not an exclusive inhibitor of Rho GTPases but also causes activation of Rho GTPases that likely assist in proinflammatory in vivo effects by the *C difficile* toxins [83].

SUMMARY

Knowledge of interactions of colonizing bacteria with GI epithelium has expanded significantly in recent years. This has helped gastroenterologists understand more precisely, not only the protective physiology of the GI tract, but also pathogenesis of infectious GI diseases. With efforts in clinical, experimental, and in vitro studies using new techniques of molecular biochemistry, more information likely will be provided about conventional and genetically engineered probiotics, their role in GI diseases, and new therapeutic applications for inflammatory bowel diseases and GI infections.

APPENDIX 1

cDNA microarray: a technique by which expression of thousands of genes can be analyzed simultaneously on a small glass or silica slide containing gene-specific DNA probes

Commensal-associated molecular patterns (CAMPS) or pathogen-associated molecular patterns (PAMPS): conserved molecular structures found on nonpathogenic microbes and pathogens but not present in higher eukaryotes

Downstream: events that occur later in a cascade of steps (eg, in signaling pathway)

Knockout: a selective inactivation of a specific gene in an otherwise normal organism

Laser-capture microdissection: a technique in which pure populations of host cells are acquired from sections of complex tissue by laser cutting

Ligand: any molecule that binds tightly and specifically to a large molecule (eg, lipopolysaccharide is a ligand for Toll-like receptor 4 [TLR4])

Pattern recognition receptors (PRRs): proteins such as TLRs and nucleotide oligomerization domains (NODDs) that are located in the plasma membrane or cytoplasm and recognize commensal- and pathogen-associated molecular patterns

Promoter: a region of DNA to which RNA polymerase binds before initiating the transcription

Real-time reverse transcription-polymerase chain reaction (real-time quantitative RT-PCR): in vitro amplification of a specific cDNA that has been copied from an mRNA molecule by reverse transcriptase. The exponential amplification of DNA is based on a series (typically 30 to 40) of temperature cycles utilizing the heat-stable DNA polymerase. In a real-time quantitative system, the amplification of DNA can be followed in real time (cycle by cycle), and the amplified product is quantitative.

Transcription: process in which a strand of DNA is used as a template for synthesis of a complementary RNA

Transcription factor: any protein that is required to initiate or regulate transcription

Wild-type: normal, nonmutant form of an organism, protein, or gene

References

[1] Haller D, Jobin C. Interaction between resident luminal bacteria and the host: can a healthy relationship turn sour? J Pediatr Gastroenterol Nutr 2004;38(2):123–36.

[2] Köhler H, McCormick BA, Walker WA. Bacterial–enterocyte cross-talk: cellular mechanisms in health and disease. J Pediatr Gastroenterol Nutr 2003;36(2):175–85.

[3] Jones NL, Sherman PM. Microbial interactions with gut epithelium. In: Walker WA, Goulet O, Kleinman RE, et al, editors. Pediatric gastrointestinal disease: pathophysiology, diagnosis, management. 4th edition. Hamilton, Canada: BC Decker Incorporated; 2004. p. 21–34.
[4] Yuan Q, Walker WA. Innate immunity of the gut: mucosal defense in health and disease. J Pediatr Gastroenterol Nutr 2004;38(5):463–73.
[5] Alonso A, Garcia-del Portillo F. Hijacking of eukaryotic functions by intracellular bacterial pathogens. Int Microbiol 2004;7(3):181–91.
[6] Kuhle V, Hensel M. Cellular microbiology of intracellular *Salmonella enterica*: functions of the type III secretion system encoded by *Salmonella* pathogenicity island 2. Cell Mol Life Sci 2004;61(22):2812–26.
[7] Schirmer J, Aktories K. Large clostridial cytotoxins: cellular biology of Rho/Ras-glucosylating toxins. Biochem Biophys Acta 2004;1673(1–2):66–74.
[8] Monack DM, Mueller A, Falkow S. Persistent bacterial infections: the interface of the pathogen and the host immune system. Nat Rev Microbiol 2004;2(9):747–65.
[9] Neish AS. Molecular aspects of intestinal-epithelial cell-bacterial interactions that determine the development of intestinal inflammation. Inflamm Bowel Dis 2004;10(2):159–68.
[10] Wheater PR, Burkitt HG, Daniels VG. Gastrointestinal tract. In: Wheater PR, Burkitt HG, Daniels VG, editors. Functional histology: a text and colour atlas. 2nd edition. Edinburg, Great Britain: Churchill Livingstone; 1987. p. 203–4.
[11] Ouellette AJ, Bevins CL. Development of innate immunity in the small intestine. In: Sanderson IR, Walker WA, editors. Development of the gastrointestinal tract. 1st edition. Hamilton, Canada: BC Decker Inc; 2000. p. 147–64.
[12] Brandtzaeg P. Development, regulation and function of secretory immunity. In: Delvin EE, Lentze MJ, editors. Gastrointestinal functions. Nestle Nutrition Workshop Series, pediatric program. Volume 46. Nestle Limited. Philadelphia: Vevey/Lippincott Williams & Wilkins; 2001. p. 91–114.
[13] Otte JM, Kiehne K, Herzig KH. Antimicrobial peptides in innate immunity of the human intestine. J Gastroenterol 2003;38(8):717–26.
[14] Ayabe T, Ashira T, Kohgo Y, et al. The role of Paneth's cells and their antimicrobial peptides in innate host defense. Trends Microbiol 2004;12(8):394–8.
[15] Mowat AMCI. Anatomical basis of tolerance and immunity to intestinal antigens. Nature Reviews 2003;3(4):331–41.
[16] Man AL, Prieto-Garcia ME, Nicoletti C. Improving M cell mediated transport across mucosal barriers: do certain bacteria hold the keys? Immunology 2004;113(1):15–20.
[17] Gebert A, Steinmetz I, Fassbender S, et al. Antigen transport in Peyer's patches: increased uptake by constant number of M cells. Am J Pathol 2004;164(1):65–72.
[18] Berg RD. The indigenous gastrointestinal microflora. Trends Microbiol 1996;4(11):430–5.
[19] Abreu MT, Arditi M. Innate immunity and toll-like receptors: clinical implications of basic science research. J Pediatr 2004;144(4):421–9.
[20] Cario E, Rosenberg IM, Brandwein SL, et al. Lipopolysaccharide activates distinct signaling pathways in intestinal epithelial cell lines expressing toll-like receptors. J Immunol 2000;164(2):966–72.
[21] Takeshita F, Leifer CA, Gursel I, et al. Cutting edge: role of toll-like receptor 9 in CpG DNA-induced activation of human cells. J Immunol 2001;167(7):3555–9.
[22] Melmed G, Thomas LS, Lee N, et al. Human intestinal epithelial cells are broadly unresponsive to toll-like receptor 2-dependent bacterial ligands: implications for host-microbial interactions in the gut. J Immunol 2003;170(3):1406–15.
[23] Otte JM, Cario E, Podolsky DK. Mechanisms of cross-hyporesponsiveness to toll-like receptor bacterial ligands in intestinal epithelial cells. Gastroenterology 2004;126(4):1054–70.
[24] Cario E, Brown D, McKee M, et al. Commensal associated molecular patterns induce selective toll-like receptor trafficking from apical membrane to cytoplasmic compartments in polarized intestinal epithelium. Am J Pathol 2002;160(1):165–73.

[25] Akira S, Takeda K. Toll-like receptor signaling. Nat Rev Immunol 2004;4(7):499–511.
[26] Gutierrez O, Pipaon C, Inohara N, et al. Induction of Nod2 in myelomonocytic and intestinal epithelial cells via nuclear factor-κB activation. J Biol Chem 2002;277(44): 41701–5.
[27] Chamaillard M, Hashimoto M, Horic Y, et al. An essential role for NOD1 in recognition of bacterial peptidoglycan containing diaminopimelic acid. Nat Immunol 2003; 4(7):702–7.
[28] Hisamatsu T, Suzuki M, Podolsky DK. Interferon-γ augments CARD4/NOD1 gene and protein expression through interferon regulatory factor-1 in intestinal epithelial cells. J Biol Chem 2003;278(35):32962–8.
[29] Inohara N, Ogura Y, Fontalba A, et al. Host recognition of bacterial muramyl dipeptide mediated through NOD2. J Biol Chem 2003;278(8):5509–12.
[30] Hugot JP, Laurent-Puig P, Gower-Rousseau C, et al. Mapping of susceptibility locus for Crohn's disease on chromosome 16. Nature 1996;379(6568):821–3.
[31] Ogura Y, Bonen DK, Inohara N, et al. A frameshift mutation in NOD2 associated with susceptibility to Crohn's disease. Nature 2001;411(6837):603–6.
[32] Lesage S, Zouali H, Cezard JP, et al. CARD15/NOD2 mutational analysis and genotype–phenotype correlation in 612 patients with inflammatory bowel disease. Am J Hum Genet 2002;70(4):845–57.
[33] Ahmad T, Armuzzi A, Bunce M, et al. The molecular classification of the clinical manifestations of Crohn's disease. Gastroenterology 2002;122(4):854–66. Erratum in: Gastroenterology 2003;125(1):281.
[34] Cuthbert AP, Fisher SA, Mirza MM, et al. The contribution of NOD2 gene mutations to the risk and site of disease in inflammatory bowel disease. Gastroenterology 2002; 122(4):867–74.
[35] Lala S, Ogura Y, Osborne C, et al. Crohn's disease and the NOD2 gene: a role for Paneth's cells. Gastroenterology 2003;125(1):47–57.
[36] Ogura Y, Lala S, Xin W, et al. Expression of NOD2 in Paneth's cells: a possible link to Crohn's ileitis. Gut 2003;52(11):1591–7.
[37] Wehkamp J, Harder J, Weichenthal M, et al. NOD2(CARD15) mutations in Crohn's disease are associated with diminished mucosal α-defensin expression. Gut 2004;53(11): 1658–64.
[38] Grimm MC, Pavli P. NOD2 mutations and Crohn's disease: are Paneth's cells and their antimicrobial peptides the link? Gut 2004;53(11):1558–60.
[39] Wehkamp J, Schmid M, Fellermann K, et al. Defensin deficiency, intestinal microbes, and the clinical phenotypes of Crohn's disease. Journal of Leukocyte Biology 2005; 77(4):460–5.
[40] Vora P, Youdim A, Thomas LS, et al. β-defensin-2 expression is regulated by TLR signaling in intestinal epithelial cells. J Immunol 2004;173(9):5398–405.
[41] Gewirtz AT, Navas TA, Lyons S, et al. Cutting edge: bacterial flagellin activates basolaterally expressed TLR5 to induce epithelial proinflammatory gene expression. J Immunol 2001;167(4):1882–5.
[42] Abreu MT, Vora P, Faure E, et al. Decreased expression of Toll-like receptor-4 and MD-2 correlates with intestinal epithelial cell protection against dysregulated proinflammatory gene expression in response to bacterial lipopolysaccharide. J Immunol 2001; 167(3):1609–16.
[43] Abreu MT, Arnold ET, Thomas LS, et al. TLR4 and MD-2 expression is regulated by immune-mediated signals in human intestinal epithelial cells. J Biol Chem 2002;277(23): 20431–7.
[44] Suzuki M, Hisamatsu T, Podolsky DK. γ-Interferon augments the intracellular pathway for lipopolysaccharide (LPS) recognition in human intestinal epithelial cells through co-ordinated up-regulation of LPS uptake and expression of the intracellular Toll-like receptor 4-MD-2 complex. Infect Immun 2003;71(6):3503–11.

[45] Hornef MW, Frisan T, Vandewalle A, et al. Toll-like receptor 4 resides in the Golgi apparatus and colocalizes with internalized lipopolysaccharide in intestinal epithelial cells. J Exp Med 2002;195(5):559–70.
[46] Hornef MW, Normark BH, Vandewalle A, et al. Intracellular recognition of lipopolysaccharide by toll-like receptor 4 in intestinal epithelial cells. J Exp Med 2003;198(8): 1225–35.
[47] Labeta MO, Vidal K, Rey-Nores JE, et al. Innate recognition of bacteria in human milk is mediated by a milk-derived highly expressed pattern recognition receptor, soluble CD14. J Exp Med 2000;191(10):1807–12.
[48] LeBouder E, Rey-Nores JE, Rushmere NK, et al. Soluble forms of toll-like receptor (TLR)2 capable of modulating TLR2 signaling are present in human plasma and breast milk. J Immunol 2003;171(10):6680–9.
[49] Claud EC, Lu L, Anton PM, et al. Developmentally regulated IκB expression in intestinal epithelium and susceptibility to flagellin-induced inflammation. Proc Natl Acad Sci U S A 2004;101(19):7404–8.
[50] Hooper LV, Xu J, Falk PG, et al. A molecular sensor that allows a gut commensal to control its nutrient foundation in a competitive ecosystem. Proc Natl Acad Sci U S A 1999; 96(17):9333–8.
[51] Hooper LV, Wong MH, Yhelin A, et al. Molecular analysis of commensal host–microbial relationship in the intestine. Science 2001;291(5505):881–4.
[52] Xu J, Bjursell MK, Himrod J, et al. A genomic view of the human–*Bacteroides thetaiotaomicron* symbiosis. Science 2003;299(5615):2074–6.
[53] Stappenbeck TS, Hooper LV, Gordon JI. Developmental regulation of intestinal angiogenesis by indigenous microbes via Paneth's cells. Proc Natl Acad Sci U S A 2002; 99(24):15451–5.
[54] Hooper LV, Stappenbeck TS, Hong CV, et al. Angiogenins: a new class of microbicidal proteins involved in innate immunity. Nat Immunol 2003;4(3):269–73.
[55] Rakoff-Nahoum S, Paglino J, Eslami-Varzaneh F, et al. Recognition of commensal microflora by Toll-like receptors is required for intestinal homeostasis. Cell 2004;118(1):229–41.
[56] Salminen S, Bouley C, Boutron-Ruault MC, et al. Functional food science and gastrointestinal physiology and function. Br J Nutr 1998;80(Suppl 1):147–71.
[57] Isolauri E, Majamaa H, Arvola T, et al. *Lactobacillus casei* strain GG reverses increased intestinal permeability induced by cow milk in suckling rats. Gastroenterology 1993; 105(6):1643–50.
[58] Rosenfeldt V, Benfeldt E, Valerius NH, et al. Effect of probiotics on gastrointestinal symptoms and small intestinal permeability in children with atopic dermatitis. J Pediatr 2004;145(5):612–6.
[59] Rosenfeldt V, Benfeldt E, Nielsen SD, et al. Effect of probiotic *Lactobacillus* strains in children with atopic dermatitis. J Allergy Clin Immunol 2003;111(2):389–95.
[60] Madsen K, Cornish A, Soper P, et al. Probiotic bacteria enhance murine and human intestinal epithelial barrier function. Gastroenterology 2001;121(3):580–91.
[61] Otte J-M, Podolsky DK. Functional modulation of enterocytes by gram-positive and gram-negative microorganisms. Am J Physiol Gastrointest Liver Physiol 2004;286(4):G613–26.
[62] Cario E, Gerken G, Podolsky DK. Toll-like receptor 2 enhances ZO-1-associated intestinal epithelial barrier integrity via protein kinase C. Gastroenterology 2004;127(1):224–38.
[63] Macpherson AJ, Uhr T. Induction of protective IgA by intestinal dendritic cells carrying commensal bacteria. Science 2004;303(5664):1662–5.
[64] Mack DR, Michail S, Wel S, et al. Probiotics inhibit enteropathogenic *E coli* adherence in vitro by inducing intestinal mucin gene expression. Am J Physiol 1999;276:G941–50.
[65] Mack DR, Ahrne S, Hyde L, et al. Extracellular MUC3 mucin secretion follows adherence of *Lactobacillus* strains to intestinal epithelial cell in vitro. Gut 2003;52(6):827–33.

[66] Wehkamp J, Harder J, Wehkamp K, et al. NF-κB- and AP-1-mediated induction of human beta defensin-2 in intestinal epithelial cells by *Escherichia coli* Nissle 1917: a novel effect of a probiotic bacterium. Infect Immun 2004;72(10):5750–8.
[67] Neish AS, Gewirtz AT, Zeng H, et al. Prokaryotic regulation of epithelial responses by inhibition of IκB-α ubiquitination. Science 2000;289(5484):1560–3.
[68] Collier-Hyams LS, Zeng H, Sun J, et al. Cutting edge: *Salmonella* avrA effector inhibits the key proinflammatory, anti-apoptotic NF-κB pathway. J Immunol 2002;169(6):2846–50.
[69] Dangl JL, Jones JD. Plant pathogens and integrated defense responses to infection. Nature 2001;411(6839):826–33.
[70] Yan F, Polk DB. Probiotic bacterium prevents cytokine-induced apoptosis in intestinal epithelial cells. J Biol Chem 2002;277(52):50959–65.
[71] Hommes DW, Peppelenbosch MP, van Deventer SJH. Mitogen-activated protein (MAP) kinase signal transduction pathways and novel anti-inflammatory targets. Gut 2003; 52(1):144–51.
[72] Ma D, Forsythe P, Bienenstock J. Live *Lactobacillus reuteri* is essential for the inhibitory effect on tumor necrosis factor alpha-induced interleukin-8 expression. Infect Immun 2004; 72(9):5308–14.
[73] Kelly D, Campbell JI, King TP, et al. Commensal anaerobic gut bacteria attenuate inflammation by regulating nuclear-cytoplasmic shuttling of PPAR-γ and RelA. Nat Immunol 2004;5(1):104–12.
[74] Jijon H, Backer J, Diaz H, et al. DNA from probiotic bacteria modulates murine and human epithelial and immune function. Gastroenterology 2004;126(5):1358–73.
[75] Dobrindt U, Hochhut B, Hentschel U, et al. Genomic islands in pathogenic and environmental microorganisms. Nat Rev Microbiol 2004;2(5):414–24.
[76] Farthing MJG. Bugs and the gut: an unstable marriage. Best Pract Res Clin Gastroenterol 2004;18(2):233–9.
[77] Steidler L, Neirynck S, Huyghebaert N, et al. Biological containment of genetically modified *Lactobacillus lactis* for intestinal delivery of human interleukin 10. Nat Biotechnol 2003;21(7):785–9.
[78] Gauthier A, Thomas NA, Finlay BB. Bacterial injection machines. J Biol Chem 2003; 278(28):25273–6.
[79] Conner SD, Schmid SL. Regulated portals of entry into the cell. Nature 2003; 422(6927):37–44.
[80] Raffatellu M, Wilson RP, Chessa D, et al. SipA, SopA, SopB, SopD, and SopE2 contribute to Salmonella enterica serotype Typhimurium invasion of epithelial cells. Inf Immunol 2005;73(1):146–54.
[81] Neish AS. Bacterial inhibition of eukaryotic proinflammatory pathways. Immunol Res 2004;29:175–86.
[82] Beaugerie L, Petit JC. Microbial–gut interactions in health and disease. Antibiotic-associated diarrhoea. Best Pract Res Clin Gastroenterol 2004;18(2):337–52.
[83] Gerhard R, Tatge H, Genth H, et al. *Clostridium difficile* toxin A induces expression of the stress-induced early gene product RhoB. J Biol Chem 2005;280(2):1499–505.

Gastroenterol Clin N Am 34 (2005) 401–412

GASTROENTEROLOGY CLINICS
OF NORTH AMERICA

Bacterial Regulation of Intestinal Immune Responses

Thomas T. MacDonald, PhD, FRCPath, FMedSci*,
John N. Gordon, MB, ChB†

Division of Infection, Inflammation and Repair, University of Southampton School of Medicine, Southampton, Hampshire, United Kingdom

As an affirmation of the importance of the gut immune system, it is often stated there is more lymphoid tissue and immune cells in the gastrointestinal tract than the rest of the body combined. It is reasonably surmised that the physiologic need to have a single layer of absorptive epithelium over the vast surface of the gut for nutrient and water absorption makes it particularly vulnerable to infection. At the same time, however, microbes have taken advantage of this niche so that all healthy individuals have a complex and extensive commensal flora, particularly in the lower small bowel and colon. Mutualism exists between the host and these bacteria, a good example of which is metabolism by the flora of indigestible substances, such as fiber into short-chain fatty acids, which provide an energy source for colonocytes.

The presence of an abundant microbial flora in the gut presents problems for the immune system. T and B cell development is designed to purge or anergize autoreactive cells so that the functional repertoire of specificities is directed against foreign antigens of infectious agents. In humans, there is extensive T and B cell development in utero in a sterile environment, but at birth there is rapid and massive colonization of the gut with commensal microbes that are antigenic and persist lifelong. One of the central issues in biology is how the gut immune system can distinguish between antigens of commensal flora and antigens of pathogens.

A second and equally important consideration is how the presence of the intestinal microbiota influences the immune system in general. This was always a consideration, and a great deal of work in the 1960s and 1970s investigated

†Dr. Gordon is supported by a fellowship from the NHS Research and Development Fund, HOPE, and the Belmont Trust.

*Corresponding author. Barts and the London School of Medicine and Dentistry, Institute of Cell and Molecular Science, 4 Newark Street, London E1 2AD, United Kingdom. *E-mail address*: t.t.macdonald@qmul.ac.uk (T.T. MacDonald).

0889-8553/05/$ – see front matter
doi:10.1016/j.gtc.2005.05.012

gastro.theclinics.com

the immune system in germ-free animals. Researchers were always aware that the microbial flora must be influencing immune responses, because components of the bacterial cell, such as lipopolysaccharide or peptidoglycan, are extremely potent modulators of in vitro and in vivo immunity. The transformation in understanding this idea mechanistically came with the discovery, in 1997, of mammalian pattern recognition receptors (PRRs), in this case, the toll-like receptors (TLRs), whose function is to recognize conserved structures on bacteria and viruses [1]. Eleven TLRs and Nod1 and Nod2, the intracellular PRRs, have been discovered (Table 1) [2–4].

PRR signaling in cells of the immune system but is particularly apropos in terms of gut immunology because this is a site where there must be constant exposure to PRR ligands. There is now abundant evidence that signaling through TLRs affects dendritic cell function, which determines T cell differentiation and antibody responses to T cell–dependent antigens [5]. This article evaluates the extent to which the products of the commensal flora regulate immune responses in the gut.

NORMAL MICROBIAL FLORA

The normal microbial flora is responsible for the abundant and highly activated mucosal immune system in healthy individuals. The human small bowel contains 200 to 300 Peyer's patches (defined as aggregates of >10 lymphoid follicles) and tens of thousands of isolated follicles [6,7]. Colonic mucosa does not contain Peyer's patches, but has many single follicles [8,9]. Peyer's patches always contain large reactive germinal centers and active T cell zones between the B cell follicles [10]. The mucosa between the follicles is also filled with

Table 1
Mammalian pattern recognition receptors and their ligands

Pattern recognition receptor	Ligand
TLR1	Bacterial lipopeptides
TLR2	Bacterial cell wall lipoteichoic acid, lipoproteins
TLR3	Bacterial cell wall peptidoglycan, double-stranded RNA
TLR4	Bacterial lipopolysaccharide
TLR5	Bacterial flagellin
TLR6	Diacylated and triacylated bacterial lipopeptides
TLR7	Guanosine-rich and uridine-rich single-stranded viral RNA
TLR8	Guanosine-rich and uridine-rich single-stranded viral RNA
TLR9	Bacterial and viral DNA
TLR10	Not known
Nod1	GM-tri-DAP (mesodiaminopimelate-containing *N*-acetylglucosamine-*N*-acetyl-ramic acid tripeptide), a peptidoglycan hydrolysis product found in gram-negative bacteria and some gram-positive bacteria
Nod2	Muramyl dipeptide in cell wall peptidoglycans from all bacteria
Peptidoglycan recognition protein	Bacterial peptidoglycan

lymphocytes in the epithelium and lamina propria [11]. There is about 1 lymphocyte for every 10 epithelial cells in the small bowel. These are mostly $CD8^+$, 90% T cell receptor $\alpha\beta$ positive, and 10% T cell receptor $\gamma\delta$ positive. In the colon, there are only 2 to 5 lymphocytes per 100 epithelial cells and a higher frequency of epithelial CD4 cells. The lamina propria, especially in the small bowel, is filled with $CD4^+$ cells and IgA plasma cells and macrophages, dendritic cells, mast cells, and the occasional eosinophil.

Analysis of the gut tissues of germ-free mice reveals a quiescent and sparse mucosal immune system [12]. Peyer's patches are present, but are small, do not contain germinal centers, and have small T cell zones (Fig. 1). The lamina propria contains essentially no IgA plasma cells or CD4 cells. Intraepithelial lymphocytes also are rare. In man, the best clue for the dependence of the mucosal immune system on gut bacteria comes from the relatively rare analyses of newborn gut. Peyer's patches at birth do not have germinal centers, and there are few IgA plasma cells in the lamina propria [13]. Mucosal T cell numbers also are low [13,14].

It is possible to reconstitute adult germ-free mice with either a single microbe or a complex flora. When this reconstitution is done, many studies have shown that it restores the mucosal immune system to the abundant and highly activated cells and tissues seen in mice colonized with bacteria from birth [15–18]. There is, therefore, no doubt that indigenous microbes are responsible

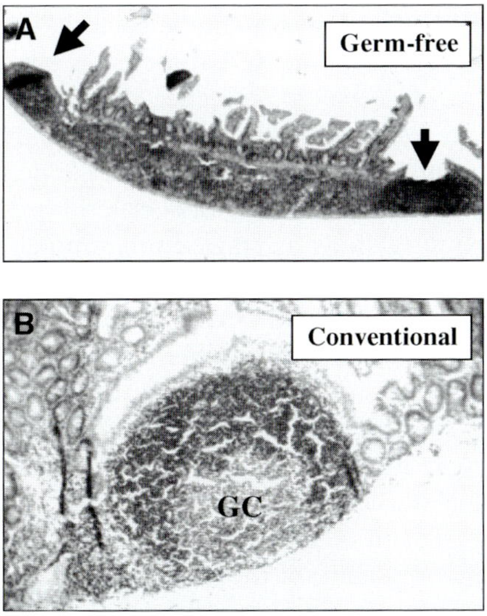

Fig. 1. The normal microbial flora is necessary for the development of gut-associated lymphoid tissue. (*A*) The histologic appearance of Peyer's patches in the ileum of a germ-free mouse. The tissue is acellular, and two primary follicles are seen (*arrows*) (hematoxylin-eosin, original magnification ×10). (*B*) A Peyer's patch from a normal mouse. The follicle contains a large germinal center (GC) (hematoxylin-eosin, original magnification ×40).

for the mucosal immune system seen in healthy individuals. It would be premature, however, to assume that all of the activated T and B cells in Peyer's patches and the mucosa represent cells responding to antigens of the flora. Mice have been created who are unable to signal through the B cell receptor for antigen (surface immunoglobulin). These animals have normal numbers of primary B cells, but cannot respond to antigen. Their Peyer's patches, but not spleen, contain large reactive germinal centers [19]. This finding suggests that much of the B cell activation in the gut is not antigen-specific, but may be driven through PRRs. Lipopolysaccharide is a well-known potent B cell mitogen. This is not to say that antigen-specific responses do not occur because it has been shown with many antigens and oral vaccines that specific IgA appears after vaccination. This response may, however, occur on a background of antigen-nonspecific activation.

Monoassociation of germ-free mice with the benign gram-negative commensal bacterium *Morganella morganii* has provided interesting insights into the relationship between the flora and the gut immune system [20]. After colonization, germinal centers appear in Peyer's patches, and a secretory IgA response is generated. After about 2 months, germinal centers disappear, but the secretory IgA response persists, presumably as a result of activation of memory B cells. When the IgA response is analyzed, 95% of the IgA produced is not specific for *M. morganii* antigens, indicating that secretory IgA responses are partly antigen-specific, but mostly polyclonal, owing to the T cell–dependent activation of B1 polyspecific cells. More recent studies have suggested that dendritic cells transport enteric bacteria from the mucosa to the mesenteric nodes, where a secretory IgA response is generated [21].

The normal flora also seems to influence the development of a diverse preimmune antibody repertoire. In humans and mice, the B cell repertoire is generated by recombinatorial joining of V, D, and J gene segments in the bone marrow, but in rabbits, this occurs in gut-associated lymphoid tissue (GALT). Rabbits with a limited flora show a reduced preimmune B cell repertoire and reduced GALT tissue [22]. Colonization of sterile rabbit appendix with different bacteria indicates that development of GALT is not antigen specific [22] because there are large variations in the ability of different commensal bacteria to restore GALT function. This situation is emphasized further by the fact that in rabbits there is a marked shift in the V region allotypes expressed by GALT B cells after birth, which is prevented by surgically removing GALT and sterilizing the appendix [23]. Detailed studies indicate that this shift may be driven by a gut bacterial superantigen. Taken together with the mouse data, it now seems clear that at least the B cell component of the mucosal immune system is driven by the flora, but in a largely antigen-nonspecific way, probably via PRRs.

For T cells, the situation is probably more complex, and the responses may be more peptide specific. There is also a possibility that superantigens from the flora may contribute to the overall level of T cell activation in the gut. The flora also may be responsible for one of the most curious features of intraepithelial lymphocytes–that in normal animals they are composed of relatively few clones of T cells [24]. In germ-free rats, the T cell repertoire in intraepithelial lymphocytes is polyclonal; however, after bacterial colonization, the repertoire changes and becomes dominated by a few clones, suggesting responses to a few immunodominant peptides or perhaps presentation of only a few lipopeptides to intraepithelial lymphocytes through the relatively nonpolymorphic class 1b MHC molecules expressed on epithelial cells [24,25]. These are also of interest because they are induced by luminal contents [26].

ROLE OF THE BACTERIAL FLORA IN REGULATING ORAL TOLERANCE

Oral tolerance is thought to be one of the mechanisms that prevents adverse responses to food proteins. At high doses, oral antigens induce apoptosis of antigen-specific T cells, but at physiological doses, oral antigens generate Th3 regulatory T cells in Peyer's patches, which then suppress immune responses by secreting immunosuppressive transforming growth factor-β or interleukin (IL)-10 [27].

The most popular antigen to study oral tolerance is ovalbumin (OVA), and numerous investigators have examined oral tolerance to OVA in germ-free animals. Early studies indicated that feeding a single dose of OVA to germ-free mice abrogated the subsequent antibody response to parenteral OVA immunization in adjuvant [28], suggesting no defect in oral tolerance in the absence of flora. In more detailed studies, feeding OVA to conventional mice induced long-term unresponsiveness of IgG and IgE responses to parenteral OVA immunization; however, in germ-free mice, feeding OVA induced short-term IgG unresponsiveness but long-term IgE unresponsiveness, suggesting that in the absence of a flora, feeding antigen did not suppress the Th2 responses on which IgE antibody production is dependent [29]. Contrasting results have been obtained, however. Vigorous IgG1 antibody responses were seen in germ-free mice given OVA orally. In terms of T cell responses, it also has been stated that germ-free mice are refractory to the induction of oral tolerance [30]. Differences also may exist in the capacity of germ-free animals to make Th1 and Th2 responses, which influences the development of IgE response. Feeding OVA to conventional mice results in Th1 and Th2 tolerance, but feeding OVA to germ-free mice results in only Th1 tolerance, and Th2 IgE responses are unaffected [31], consistent with previous work [29]. Neonatal colonization of germ-free mice with bacteria results in animals in whom Th1 and Th2 responses can be suppressed by antigen feeding. Neonatal colonization with bacteria also seems critical in the ability of *Escherichia coli* enterotoxin to ablate oral tolerance [32].

It was shown many years ago that germ-free rats had abundant IgE-positive B cells in their Peyer's patches [33]. This observation is consistent with the idea that in the absence of a commensal flora, neonatal Th2 to Th1 switching does not occur, predisposing to IgE responses and allergy. It is highly likely that this Th2 to Th1 switch is driven by the commensal flora signaling through TLRs to antigen-presenting cells in the gut and inducing Th1 skewing cytokines, such as IL-12 or IL-23. Compelling evidence for an important relationship between IgE and signaling from the flora comes from studies of animals defective in TLR4, the receptor for lipopolysaccharide [34]. Feeding peanut allergen with a mucosal adjuvant induced high levels of IgE in lipopolysaccharide-unresponsive mice and a Th2-polarized cytokine response. If synthetic TLR9 ligands are given at the same time as oral sensitization, IgE responses are reduced. Finally, antibiotic decontamination of the gut results in higher IgE responses [34]. Taken together, these results indicate that the normal microbial flora is a necessary prerequisite before oral administration of antigens can suppress Th2 responses and IgE.

INTERACTIONS OF THE MICROBIAL FLORA WITH DENDRITIC CELLS

PRRs are expressed on virtually all cell types in the gut, so there is potential for extensive interaction between the flora and the host. The overall importance of this topic is well shown by the fact that the flora induces cytoprotective heat-shock proteins in epithelial cells [35] and that in the functional absence of TLR, the mouse gut is susceptible to dextran sodium sulfate–induced colitis [36]. The present article concentrates on the more immunologic aspects of this interaction. However, epithelial cell biology and immune cell biology are inextricably entwined in the gut.

Follicle-associated epithelium (FAE) of Peyer's patches constitutively secretes the chemokines CCL9 and CCL20 [37,38]. Dendritic cells express the known receptors of CCL20 and CCR6, and mice lacking CCR6 do not have dendritic cells in the dome region of their Peyer's patches [39,40]. Likewise, blocking CCL9 reduces CD11b-positive dendritic cells in the subepithelial dome [41]. The pathways that control chemokine expression by FAE are not well known, but FAE CCL20 expression seems to be driven by inducer $CD4^+$, $CD3^-$ cells in the Peyer's patch anlage, which signal to epithelium through lymphotoxin $\alpha 1\beta 2$ on the Peyer's patch inducer cells and the LTβ receptor on epithelium [42]. FAE CCL20 expression is normal in Myd88 null mice, who are unable to signal through TLRs [42]. A role for the flora should not be discounted, however, because M cells in the FAE phagocytose bacterial components from the lumen, and intracellular Nod1/Nod2 may be activated, which could induce chemokine expression in a Myd88-independent fashion. Likewise, M cells also express the secreted peptidoglycan recognition protein (PGRP-S), and other cells in the FAE express the transmembrane form of PGRP-L [43]. Mice lacking PGRP-L are not particularly susceptible to infection, but PGRP-L may play a more specialized role in the FAE of the gut [44]. There also are some differences in Peyer's patch dendritic cells between conventional and germ-free rats. In the germ-free state, the FAE of the Peyer's patch contains immature dendritic cells and few lymphocytes. After colonization with bacteria, these immature dendritic cells disappear from the FAE [45].

The Peyer's patches and colonic lymphoid follicles are also specialized sites where M cells transport luminal antigens and TLR ligands derived from the flora into the subepithelial dome. Although FAE attracts dendritic cells to this site, their function is probably determined by signals coming from the lumen via M cells (Fig. 2). In mice, the subepithelial dome region of Peyer's patches contains three populations of dendritic cells: cells expressing CD11b, cells expressing CD8αα, and a double-negative subset [46,47]. In humans, subepithelial dendritic cells express CD11c and probably are myeloid dendritic cells [48]. Some of these cells also express DC-SIGN, a coreceptor for human immunodeficiency virus [48]. Little functional work has been done on these cells. In mice, the Peyer's patch CD11b-positive dendritic cells seem to make IL-10 and may skew T cells along a regulatory pathway [47]. In children, Peyer's patch dendritic cells in the subepithelial dome express IL-12, and T cells isolated from human Peyer's patches secrete high amounts of interferon-γ, which can be blocked with anti-IL-12 [49].

A more recent development has been the observation that lamina propria dendritic cells reach through the epithelium and can capture commensal and pathogenic microbes [50–52]. It is assumed that these dendritic cells then

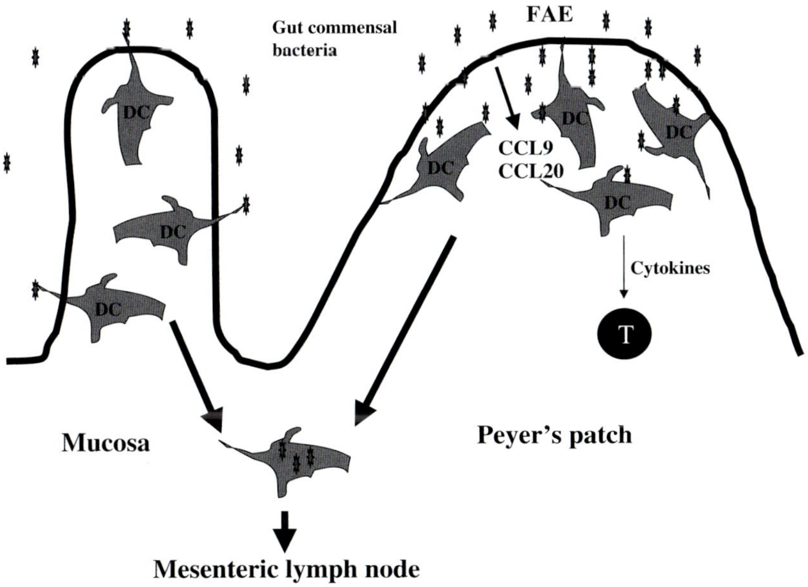

Fig. 2. The intimate association between dendritic cells (DCs) and the gut epithelium. In the villi, DCs send processes through into the lumen and capture commensal bacteria. It is highly likely that they then migrate to the mesenteric node and initiate a secretory IgA response. In the Peyer's patch, DCs in the subepithelial dome region send processes into the follicle-associated epithelium (FAE). DCs are attracted to this region because the FAE produces chemokines, which attract DC precursors. At this site, however, they are exposed to bacteria and bacterial products transported into the dome by M cells. These interact with the DCs through toll-like receptors, and the cytokines they produce determine whether activated T cells become Th1, Th2, or T regulatory cells.

migrate to the mesenteric lymph nodes, where they present bacterial antigens to T and B cells (see Fig. 2).

The potential of luminal organisms to skew dendritic cell function has been studied extensively in vitro, largely focusing on probiotics, but lactobacilli and bifidobacteria are also members of the commensal flora. Generally, interaction of bacteria with dendritic cells results in a maturation signal, with the dendritic cell changing from an immature, highly endocytic cell expressing low levels of accessory molecules with mostly intracellular MHC molecules into a mature cell expressing high levels of accessory molecules and high levels of surface MHC [5].

It is not known how the flora may alter dendritic cell function in the gut. The addition of different *Lactobacillus* strains to mouse bone marrow–derived dendritic cells induces cytokine production, but at physiologic doses; more IL-12, tumor necrosis factor (TNF)-α, and IL-6 are produced than IL-10 [53], which would suggest that probiotics induce proinflammatory cytokine responses. Addition of *Streptococcus pyogenes* to human monocyte-derived dendritic cells increases accessory molecules and IL-12, IL-23, and IL-27 [54]. In a similar vein, addition of *E coli* to monocyte-derived dendritic cells stimulates the production of IL-12, TNF-α, IL-6, and IL-10, especially if interferon-γ is also added [55]. Quantitatively, more IL-12, TNF-α, and IL-6 are produced than IL-10 [55]. In a broad analysis of the capacity of gram-positive and gram-negative commensals from the human gut to induce IL-12, IL-23, and IL-27, Smits et al [56] showed that only gram-negative organisms primed dendritic cells to induce Th1 responses. VSL#3 is a complex probiotic containing four *Lactobacillus* species, three *Bifidobacterium* species, and *Streptococcus salivarius*. Addition of VSL#3 to have bone marrow-derived dendritic cells induces IL-10 [57]. It also has been reported that VSL#3 induces IL-10 in human blood and gut dendritic cells [58].

It is difficult to draw definitive conclusions from these often disparate studies, but if one excludes probiotics, it would seem clear that gut gram-negative and probably gram-positive bacteria can drive dendritic cells to produce cytokines that skew T cells along a Th1 pathway. This situation might explain the Th1 dominance of the human mucosal immune system [49,59], but does raise the issue of whether individuals who develop food allergies might be skewed along a different direction by different members of the flora. There is evidence that atopic children have a different flora compared with nonatopic children [60].

SUMMARY

There is now little doubt that what is considered the mucosal immune system of healthy individuals is a host response to the commensal bacterial flora. At a more detailed level, it is becoming clear that the presence of the flora must alter epithelial and immune function in the gut, and that immune responses to nominal antigens occur in an environment where products of the flora are controlling dendritic cell function. Unraveling the cross-talk between the commensal flora and the mucosal immune system is likely to give insight into why tolerance develops in most cow's milk–allergic infants and why inappropriate T cell responses to the flora lead to Crohn's disease.

References

[1] Medzhitov R, Preston-Hurlburt P, Janeway CA Jr. A human homologue of the *Drosophila* Toll protein signals activation of adaptive immunity. Nature 1997;388:394–7.
[2] Philpott DJ, Girardin SE. The role of Toll-like receptors and Nod proteins in bacterial infection. Mol Immunol 2004;41:1099–108.
[3] Chamaillard M, Inohara N, Nunez G. Battling enteroinvasive bacteria: Nod1 comes to the rescue. Trends Microbiol 2004;12:529–32.
[4] Pasare C, Medzhitov R. Toll-like receptors: linking innate and adaptive immunity. Microbes Infect 2004;6:1382–7.
[5] Iwasaki A, Medzhitov R. Toll-like receptor control of the adaptive immune responses. Nat Immunol 2004;5:987–95.
[6] Cornes JS. Number, size, and distribution of Peyer's patches in the human small intestine. Gut 1965;6:225–33.
[7] Van Kruiningen HJ, West AB, Freda BJ, Holmes KA. Distribution of Peyer's patches in the distal ileum. Inflamm Bowel Dis 2002;8:180–5.
[8] Dukes C, Bussey HJF. The number of lymphoid follicles of the human large intestine. J Pathol Bacteriol 1925;111–6.
[9] Langman JM, Rowland R. The number and distribution of lymphoid follicles in the human large intestine. J Anat 1986;149:189–94.
[10] Spencer J, Finn T, Isaacson PG. Human Peyer's patches: an immunohistochemical study. Gut 1986;27:405–10.
[11] MacDonald TT. Mucosal lymphoid tissue. In: Challacombe SJ, Brostoff J, editors. Food allergy and intolerance. London: Bailliere Tindall; 2002. p. 3–13.
[12] Pickard K, Bremner R, Gordon J, MacDonald T. Host responses to the normal gut imcroflora: immune responses. Best Pract Res Clin Gastroenterol 2004;18:271–85.
[13] Spencer J, MacDonald T. Development of the human mucosal immune system. In: MacDonald TT, editor. Ontogeny of the intestinal immune system. Boca Raton (FL): CRC Press; 1990. p. 23–50.
[14] Ormala T, Rintala R, Savilahti E. T-cells and HLA-class II expression in the large intestine of infants in the early postnatal period. J Pediatr Gastroenterol Nutr 1998;26:422–8.
[15] Crabbe PA, Bazin H, Eyssen H, Heremans JF. The normal microbial flora as a major stimulus for proliferation of plasma cells synthesizing IgA in the gut: the germ-free intestinal tract. Int Arch Allergy Appl Immunol 1968;34:362–75.
[16] Moreau MC, Ducluzeau R, Guy-Grand D, Muller MC. Increase in the population of duodenal immunoglobulin A plasmocytes in axenic mice associated with different living or dead bacterial strains of intestinal origin. Infect Immun 1978;21:532–9.
[17] Klaasen HL, Van der Heijden PJ, Stok W, et al. Apathogenic, intestinal, segmented, filamentous bacteria stimulate the mucosal immune system of mice. Infect Immun 1993;61:303–6.
[18] Umesaki Y, Okada Y, Matsumoto S, Imaoka A, Setoyama H. Segmented filamentous bacteria are indigenous intestinal bacteria that activate intraepithelial lymphocytes and induce MHC class II molecules and fucosyl asialo GM1 glycolipids on the small intestinal epithelial cells in the ex-germ-free mouse. Microbiol Immunol 1995;39:555–62.
[19] Casola S, Otipoby KL, Alimzhanov M, et al. B cell receptor signal strength determines B cell fate. Nat Immunol 2004;5:317–27.
[20] Jiang HQ, Thurnheer MC, Zuercher AW, Boiko NV, Bos NA, Cebra JJ. Interactions of commensal gut microbes with subsets of B- and T-cells in the murine host. Vaccine 2004;22:805–11.

[21] Macpherson AJ, Uhr T. Induction of protective IgA by intestinal dendritic cells carrying commensal bacteria. Science 2004;303:1662–5.
[22] Rhee KJ, Sethupathi P, Driks A, Lanning DK, Knight KL. Role of commensal bacteria in development of gut-associated lymphoid tissues and preimmune antibody repertoire. J Immunol 2004;172:1118–24.
[23] Rhee KJ, Jasper PJ, Sethupathi P, Shanmugam M, Lanning D, Knight KL. Positive selection of the peripheral B cell repertoire in gut-associated lymphoid tissues. J Exp Med 2005;201:55–62.
[24] Balk SP, Ebert EC, Blumenthal RL, et al. Oligoclonal expansion and CD1 recognition by human intestinal intraepithelial lymphocytes. Science 1991;253:1411–5.
[25] Helgeland L, Dissen E, Dai KZ, Midtvedt T, Brandtzaeg P, Vaage JT. Microbial colonization induces oligoclonal expansions of intraepithelial CD8 T cells in the gut. Eur J Immunol 2004;34:3389–400.
[26] Colgan SP, Pitman RS, Nagaishi T, et al. Intestinal heat shock protein 110 regulates expression of CD1d on intestinal epithelial cells. J Clin Invest 2003;112:745–54.
[27] Weiner HL. Induction and mechanism of action of transforming growth factor-beta-secreting Th3 regulatory cells. Immunol Rev 2001;182:207–14.
[28] Moreau MC, Corthier G. Effect of the gastrointestinal microflora on induction and maintenance of oral tolerance to ovalbumin in C3H/HeJ mice. Infect Immun 1988; 56:2766–8.
[29] Moreau MC, Gaboriau-Routhiau V. The absence of gut flora, the doses of antigen ingested and aging affect the long-term peripheral tolerance induced by ovalbumin feeding in mice. Res Immunol 1996;147:49–59.
[30] Maeda Y, Noda S, Tanaka K, et al. The failure of oral tolerance induction is functionally coupled to the absence of T cells in Peyer's patches under germfree conditions. Immunobiology 2001;204:442–57.
[31] Sudo N, Sawamura S, Tanaka K, Aiba Y, Kubo C, Koga Y. The requirement of intestinal bacterial flora for the development of an IgE production system fully susceptible to oral tolerance induction. J Immunol 1997;159:1739–45.
[32] Gaboriau-Routhiau V, Raibaud P, Dubuquoy C, Moreau MC. Colonization of gnotobiotic mice with human gut microflora at birth protects against *Escherichia coli* heat-labile enterotoxin-mediated abrogation of oral tolerance. Pediatr Res 2003;54:739–46.
[33] Durkin HG, Bazin H, Waksman BH. Origin and fate of IgE-bearing lymphocytes: I. Peyer's patches as differentiation site of cells simultaneously bearing IgA and IgE. J Exp Med 1981;154:640–8.
[34] Bashir ME, Louie S, Shi HN, Nagler-Anderson C. Toll-like receptor 4 signaling by intestinal microbes influences susceptibility to food allergy. J Immunol 2004;172:6978–87.
[35] Kojima K, Musch MW, Ren H, et al. Enteric flora and lymphocyte-derived cytokines determine expression of heat shock proteins in mouse colonic epithelial cells. Gastroenterology 2003;124:1395–407.
[36] Rakoff-Nahoum S, Paglino J, Eslami-Varzaneh F, Edberg S, Medzhitov R. Recognition of commensal microflora by toll-like receptors is required for intestinal homeostasis. Cell 2004;118:229–41.
[37] Rossi DL, Vicari AP, Franz-Bacon K, McClanahan TK, Zlotnik A. Identification through bioinformatics of two new macrophage proinflammatory human chemokines: MIP-3alpha and MIP-3beta. J Immunol 1997;158:1033–6.
[38] Tanaka Y, Imai T, Baba M, et al. Selective expression of liver and activation-regulated chemokine (LARC) in intestinal epithelium in mice and humans. Eur J Immunol 1999; 29:633–42.
[39] Cook DN, Prosser DM, Forster R, et al. CCR6 mediates dendritic cell localization, lymphocyte homeostasis, and immune responses in mucosal tissue. Immunity 2000; 12:495–503.

[40] Varona R, Villares R, Carramolino L, et al. CCR6-deficient mice have impaired leukocyte homeostasis and altered contact hypersensitivity and delayed-type hypersensitivity responses. J Clin Invest 2001;107:R37–45.
[41] Zhao X, Sato A, Dela Cruz CS, et al. CCL9 is secreted by the follicle-associated epithelium and recruits dome region Peyer's patch CD11b + dendritic cells. J Immunol 2003; 171:2797–803.
[42] Rumbo M, Sierro F, Debard N, Kraehenbuhl JP, Finke D. Lymphotoxin beta receptor signaling induces the chemokine CCL20 in intestinal epithelium. Gastroenterology 2004;127:213–23.
[43] Lo D, Tynan W, Dickerson J, et al. Peptidoglycan recognition protein expression in mouse Peyer's patch follicle associated epithelium suggests functional specialization. Cell Immunol 2003;224:8–16.
[44] Choe KM, Lee H, Anderson KV. *Drosophila* peptidoglycan recognition protein LC (PGRP-LC) acts as a signal-transducing innate immune receptor. Proc Natl Acad Sci U S A 2005;102: 1122–6.
[45] Yamanaka T, Helgeland L, Farstad IN, Fukushima H, Midtvedt T, Brandtzaeg P. Microbial colonization drives lymphocyte accumulation and differentiation in the follicle-associated epithelium of Peyer's patches. J Immunol 2003;170:816–22.
[46] Iwasaki A, Kelsall BL. Localization of distinct Peyer's patch dendritic cell subsets and their recruitment by chemokines macrophage inflammatory protein (MIP)-3alpha, MIP-3beta, and secondary lymphoid organ chemokine. J Exp Med 2000;191:1381–94.
[47] Iwasaki A, Kelsall BL. Unique functions of CD11b +, CD8 alpha +, and double-negative Peyer's patch dendritic cells. J Immunol 2001;166:4884–90.
[48] Jameson B, Baribaud F, Pohlmann S, et al. Expression of DC-SIGN by dendritic cells of intestinal and genital mucosae in humans and rhesus macaques. J Virol 2002;76:1866–75.
[49] MacDonald T, Monteleone G. Human gut-associated lymphoid tissue. In: Mestecky J, Lamm ME, Strober W, et al, editors. Mucosal immunology 3. San Diego: Academic Press; 2004. p. 407–13.
[50] Rescigno M, Urbano M, Valzasina B, et al. Dendritic cells express tight junction proteins and penetrate gut epithelial monolayers to sample bacteria. Nat Immunol 2001;2:361–7.
[51] Becker C, Wirtz S, Blessing M, et al. Constitutive p40 promoter activation and IL-23 production in the terminal ileum mediated by dendritic cells. J Clin Invest 2003;112: 693–706.
[52] Niess JH, Brand S, Gu X, et al. CX3CR1-mediated dendritic cell access to the intestinal lumen and bacterial clearance. Science 2005;307:254–8.
[53] Christensen HR, Frokiaer H, Pestka JJ. Lactobacilli differentially modulate expression of cytokines and maturation surface markers in murine dendritic cells. J Immunol 2002;168:171–8.
[54] Veckman V, Miettinen M, Pirhonen J, Siren J, Matikainen S, Julkunen I. *Streptococcus pyogenes* and *Lactobacillus rhamnosus* differentially induce maturation and production of Th1-type cytokines and chemokines in human monocyte-derived dendritic cells. J Leukoc Biol 2004;75:764–71.
[55] Karlsson H, Larsson P, Wold AE, Rudin A. Pattern of cytokine responses to gram-positive and gram-negative commensal bacteria is profoundly changed when monocytes differentiate into dendritic cells. Infect Immun 2004;72:2671–8.
[56] Smits HH, van Beelen AJ, Hessle C, et al. Commensal gram-negative bacteria prime human dendritic cells for enhanced IL-23 and IL-27 expression and enhanced Th1 development. Eur J Immunol 2004;34:1371–80.
[57] Drakes M, Blanchard T, Czinn S. Bacterial probiotic modulation of dendritic cells. Infect Immun 2004;72:3299–309.
[58] Hart AL, Lammers K, Brigidi P, et al. Modulation of human dendritic cell phenotype and function by probiotic bacteria. Gut 2004;53:1602–9.

[59] Fuss IJ, Neurath M, Boirivant M, et al. Disparate CD4 + lamina propria (LP) lymphokine secretion profiles in inflammatory bowel disease: Crohn's disease LP cells manifest increased secretion of IFN-gamma, whereas ulcerative colitis LP cells manifest increased secretion of IL-5. J Immunol 1996;157:1261–70.
[60] Bjorksten B, Sepp E, Julge K, Voor T, Mikelsaar M. Allergy development and the intestinal microflora during the first year of life. J Allergy Clin Immunol 2001;108:516–20.

Gastroenterol Clin N Am 34 (2005) 413–436

GASTROENTEROLOGY CLINICS
OF NORTH AMERICA

Bioecologic Control of the Gastrointestinal Tract: The Role of Flora and Supplemented Probiotics and Synbiotics

Stig Bengmark, MD, PhD[a,b,*]

[a]Departments of Hepatology and Surgery, Institute of Hepatology, University College, London Medical School, 69-75 Chenies Mews, London, WC1E 6HX, United Kingdom
[b]Lund University, Lund, Sweden

EPIDEMIC OF CHRONIC DISEASES

Although in certain parts of the world acute infectious diseases still dominate as causes of death, especially in children, globally chronic diseases constitute the leading causes of morbidity and mortality today. The World Health Organization estimates that 46% of global disease burden and 59% of global mortality are due to chronic disease; 35 million individuals die each year from chronic diseases, and this statistic has been increasing steadily. The increase, which seems to have its beginning at the time of the Industrial Revolution (eg, the mid-1850s), was relatively slow during the first 100 years, but during recent decades the increase in morbidity and mortality has obtained epidemic proportions. Circumstantial evidence supports an association of chronic disease with the transition from natural unprocessed foods to processed and often calorie-condensed foods. The correlation between increases in chronic diseases and reduction in intake of plant fibers and plant antioxidants with increase in consumption of refined sugars is obvious; the individual consumption of sugar has increased from about 1 lb/person/y in 1850 to about 100 lb/person/y in 2000. Today the fastest increase in chronic disease seems to be in the Third World. These diseases are imported to Third-World countries with an enormous surplus of cheap agricultural products, including grains, especially wheat, and dairy products, especially butter. Little consideration is given to the fact that a large proportion of individuals in these parts of the world are gluten intolerant or to the detrimental effect of such imports on local production of health-promoting fresh fruits and vegetables.

*185 Barrier Point Road, Royal Docks, London, E16 2Se, United Kingdom. *E-mail address*: s.bengmark@ucl.ac.uk

0889-8553/05/$ – see front matter
doi:10.1016/j.gtc.2005.05.002

gastro.theclinics.com

FOOD CONNECTION

The modern Western lifestyle is characterized not only by significant alterations in food consumption. Stress; lack of physical exercise; use of alcohol, tobacco, and pharmaceuticals, and increasing exposure to environmental chemicals also seem to contribute to the burden of chronic disease in Western society. Foods that are consumed–refined and calorie-condensed food products–contain large amounts of saturated and trans fatty acids, sugar and starch, and bioactive peptides such as gluten and are low in omega-3 polyunsaturated fatty acids, plant antioxidants, and health-promoting plant fibers and bacteria. Common to most of the above-mentioned food ingredients is that they affect the function of the innate immune system, the inflammatory response, and the individual's resistance to disease [1]. Plant fibers, antioxidants, and, to some extent, polyunsaturated fatty acids enforce the resistance to disease, whereas saturated and trans fatty acids, sugar and starch, peptides such as gluten, and many chemicals and pharmaceuticals, including antibiotics, suppress the resistance to disease. Consequently, most patients with chronic disease have increased acute and chronic phase response, increased inflammation/superinflammation, and metabolic syndrome [1]. Saturated fat and trans fatty acids induce significant alterations in the immune response [2]; inhibit the macrophage functions [3]; stimulate the Th2 response relative to the Th1 response; and increase the risk of getting chronic diseases, such as diabetes, certain cancers, and rheumatoid arthritis [3]. Antibiotics result in suppression of the various macrophage functions, including chemiluminescence response, chemotactic motility, bactericidal and cytostatic ability, and lymphocyte proliferation [4,5].

The links between chronic diseases of the gastrointestinal tract, especially inflammatory bowel disease (IBD) and irritable bowel syndrome (IBS), and chronic phase response and metabolic syndrome have not been studied to the extent they should. Elevated levels of proinflammatory cytokines, increased coagulability, and signs of epithelial and endothelial dysfunction, changes that significantly relate to disease activity, are observed repeatedly in IBD [6,7], especially in ulcerative colitis (UC). A few studies also report insulin resistance in IBD patients [8]. Its role is not clear, but it is reported that insulin sensitivity improves when disease goes into remission.

The intolerance to various foods and their role as risk factors in various digestive tract chronic diseases have not been studied to any larger extent. When the British gastroenterologist Hunter studied food intolerance in IBS in the 1980s, he reported grains, particularly wheat (60% of patients) and corn (34% of patients); dairy products, particularly cow's milk (44% of patients) and cheese (39% of patients); and coffee (33% of patients) were the most common foods to induce intolerance [9]. Although less well studied, the pattern seems to be similar in other digestive tract chronic diseases. Cow's milk is strongly associated with high blood levels of growth factors, such as insulin-like growth factor-1, and with certain chronic diseases, such as breast cancer and prostate cancer [10,11]. Gluten-containing grains and dairy products contain morphine-like molecules, which destabilize cellular membranes in the body and induce dysfunction of the immune cells, especially in individuals with the specific genetic marker HLA-B8, which is said to exist in 30% of Western populations [12]. These individuals all seem to suffer "a leaky gut," even if only a small fraction of them develop full-blown celiac disease. The ability of human gut to digest small peptides is limited and depends much on the presence and composition of flora. Only a few lactic acid bacteria (LAB) have the ability to disintegrate these peptides. In some individuals, but not all, the liver possesses individual enzymes that eliminate all peptides before entering the general circulation. In this connection, it has been shown that long-term fermentation of dough totally eliminates gluten to a degree that the bread produced with long-term fermentation is tolerated by individuals with pronounced gluten intolerance [13]. This observation gives a new dimension to the Paleolithic diet. Early humans did not consume grains or cow's milk, and most of the food they ate was vigorously fermented, as it was commonly stored in the soil for days, weeks, and months. Early humans consumed considerably less salt, fat, and sugar, but consumed about twice as much minerals, 10 times more plant fibers, greater than 20 times more antioxidants, greater than 50 times more omega-3 fatty acids, and billions of times more live bacteria [14].

MICROBE CONNECTION

Studies in animals, including germ-free animals, and in humans show a series of benefits from living in symbiosis with microbes. Life would be difficult to sustain without access to plants and microbes. The human body, similar to that of all other animals, is "self-cleaning." It is for that purpose equipped with a "layer" of microbes at all surfaces to the exterior world: large intestine, 1 to 2 kg; skin, 200 g; oral cavity, lung, and vagina, each 20 g; nose, 10 g; and eye, 1 g [15]. The human body contains 10 to 20 times more prokaryotic cells (10^{14}) than eukaryotic cells (10^{13}). The flora is suggested to contain 30 times more genes as the rest of the body–greater than 2 million prokaryotic genes compared with 65,000 karyotic genes [16]. The beneficial/probiotic bacteria seem to tolerate poorly the Western lifestyle with its stress and poor eating habits. The intestinal mucosa is said to contain more nerve endings than any other tissue in the body, and when the individual is in stress, release of norepinephrine into the intestinal lumen reduces the good bacteria and changes the phenotype of the potentially pathogenic microorganisms, which become considerably more virulent and sometimes life-threatening [17]. Studies in animals with induced disease, such as pancreatitis, show that the preventive bacteria disappear after 4 to 6 hours, and overgrowth of pathogens and microbial translocation occurs about 6 hours after induction of disease [18,19]. Studies in critically ill humans indicate that the entire LAB flora is lost in patients after a short stay in an intensive care unit [20]. Astronauts have been shown on return to earth to have an absent or much reduced LAB flora and clear signs of overgrowth with pathogens [21].

Westerners in general have a much reduced flora, in diversity of microbial strains and number of bacteria, compared with people who live under more rural conditions; Swedish children show a much reduced flora compared with Estonian [22] and Parkistani [23] children. A United States study from 1983 showed that *Lactobacillus plantarum,* a dominating LAB among plant eaters, is found in only approximately 25% of omnivorous Americans and in about two thirds of vegetarian Americans [24], and nothing supports an assumption that this has changed to the better. A more recent Scandinavian study suggested that even in healthy individuals the most common colonic LAB are present in only about 50% or less: *Lactobacillus plantarum* in 52%, *Lactobacillus rhamnosus* in 26%, and *Lactobacillus paracasei* subsp *paracasei* 17% [25].

Only a minority of the intestinal flora is cultivable, perhaps 30%. There is a "large dark hole" about which little is known. It is possible that specific, yet unknown microbes exist, which in the future might be recognized as responsible for "unexplained" diseases such as chronic diseases of the gut, rheumatic arthritis, and autism. The increasing understanding of inflammation and function of acute phase proteins has shown that many chronic diseases in the past labeled as degenerative are actually inflammatory. More recent availability of molecular biology techniques and the understanding of the function of nanomolecules have made it possible to associate some diseases with specific microbial infections–not only chronic gastritis with *Helicobacter pylori* and arteriosclerosis with *Campylobacter* and *Helicobacter* infections, but also a range of other diseases [26].

In the 1990s, the authors' group showed a significant reduction in numbers of bacteria in patients with IBD [27]. A more recent study using modern molecular biology techniques showed a significant reduction in diversity of microbial strains in IBD patients: in UC with approximately 30% and in Crohn's disease (CD) with 50% [28]. It also has been shown that especially patients with active UC have lost the ability of holding back the potentially pathogenic flora from the intimate mucosal surface [29], and it was shown that patients with CD have a high prevalence of adherent-invasive *Escherichia coli* on the intestinal mucosa; adherent-invasive *E coli* strains were found in 22% of patients with chronic lesions of CD versus 6% of controls and 36% of patients in the so-called neoterminal ileum (eg, the last 10 cm of ileum before an ileocolic anastomosis) [30].

ANTIOXIDANT CONNECTION

Increased levels of pro-oxidants, such as homocysteine, and low levels of vitamins and key antioxidants, such as folic acid and glutathione, often are observed in patients with all types of chronic disease. It has been suggested that IBD results from an imbalance between pro-oxidant and antioxidant mechanisms [31,32]. Glutathione, a tripeptide and key intracellular antioxidant, has generated special interest [33]. Low levels of glutathione especially have been reported from studies of colonic biopsy specimens from patients with CD [34]. Vitamins and antioxidants often are given as supplements to patients with chronic gastrointestinal disease, but, to the authors' knowledge, no controlled study with systematic supply of antioxidants is reported in the literature. Several such studies have been performed in experimental animals, however. Superantioxidants (eg, antioxidants supposedly >10 times stronger in effect than vitamins such as vitamin C and E), including resveratrol (rich in red wine and peanuts), quercetin (apple, onion, tea), epigallocatechin gallate (green tee), lycopene (tomato), and curcumin (turmeric), all have the potential alone or in combination to provide health benefits. Special attention has been given more recently to curcumin, an ingredient in the spice turmeric obtained from the rhizomes of the plant *Curcuma longa Linn,* which contains 1% to 15% of the active substance. Similar to all other plant-borne antioxidants, curcumin most likely is released and absorbed after microbial fermentation in the large intestine. In experimental studies, curcumin has shown strong anti-inflammatory effects and capacity to prevent various chronic diseases, including neurodegenerative diseases such as Alzheimer's disease and various cancers. Significant prevention of digestive tract disease development has been shown in animals with liver injuries induced by carbon tetrachloride [35] and alcohol [36] and in intestinal mucosa and colitis development when induced by trinitrobenzone sulphonic acid [37,38]. Therapeutic effects, which by no means are inferior to what has been observed with other therapeutic modalities tried in experimental studies, including that of probiotics, also have been shown.

FIBER CONNECTION

There is little to support a pathogenic connection between digestive tract chronic diseases and reduced intake of plant fibers per se, even if chronic diseases in general are associated with unsatisfactory consumption of fresh fruits and vegetables, nonalcohol steatosis hepatis and cryptogenic cirrhosis suggested as possible exceptions. Efforts to supplement various plant fibers to patients with digestive tract chronic diseases most often have been unsuccessful. Some more recent studies have led to results, however, that might change clinicians' attitudes. The ability to maintain remission in UC patients through daily supplementation with 10 g of *Plantago ovata* seeds (also called psyllium or Ispaghula husk) alone or in combination with 500 mg of mesalamine was compared with daily treatment with 500 mg of mesalamine only [39]. Twelve months of treatment showed no statistical support in favor of any of the groups, a result that seemingly favors the use of psyllium because of its cheaper price and lack of side effects. Daily supply of 30 g of germinated barley foodstuff, a by-product from breweries, rich in hemicellulose and in glutamine, was tried in 39 patients with mild-to-moderate active UC [40] and shown to increase significantly numbers of LAB in stool, especially of *Bifidobacterium* and *Eubacterium*; reduce the disease activity; and increase concentration of short-chain fatty acids, a key energy source for colonic mucosa, essential for mucosal growth. Short-chain fatty acids, especially butyrate, are shown significantly to inhibit nuclear factor-κB activation of lamina propria macrophages, reduce the number of neutrophils in crypts and surface epithelia, and reduce the density of lamina propria lymphocytes/plasma cells in patients with UC [41]. By-products from breweries are rich in various important nutrients, including large amounts of vitamins such as various B vitamins, and the effects observed also can be linked with several such ingredients or a combination of ingredients. A controlled study with oat bran as the fiber source was conducted in 22 supplemented and 10 control patients with quiescent UC. Daily supplementation during 3 months of 60 g of oat bran (equivalent to 20 g of dietary fiber) resulted in a significant increase in fecal butyrate (average 36%) and significant reduction of abdominal pain. All UC patients tolerated this large dose of fiber well, and signs of relapse were seen in none of the colitis patients [42]. Another controlled study reported significant reduction in inflammation of the mucosa of the ileal reservoir on endoscopy and histology after 3 weeks of daily supplementation of 24 g of inulin to patients with ileal pouch–anal anastomosis [43]. More recent studies also suggest that supplementation of nondigestible oligosaccharides per se enhances bacterial colonization resistance against *Clostridium difficile,* at least in vitro [44]. Nonabsorbable disaccharides, such as lactulose, and plant fibers, such as psyllium, also have proved effective to decrease blood ammonia and reduce encephalopathy in patients with liver cirrhosis [45]. Daily supplementation of about 5 g/d of pectin or about two unripe bananas (rich in pectin and cellulose) was proven extremely effective as supplement to rehydration against persistent diarrhea in children in Bangladesh. Recovery on the third day was seen in 59% in the green banana group and 55% in the pectin group compared with 15% in a rice-only group [46]. Studies also suggest significant reduction in encephalopathy and improvement of liver function in patients with chronic liver disease when given a daily supplement of a combination of 10 g of four fibers (β-betaglucan, inulin, pectin, resistant starch) [47].

PREBIOTICS, PROBIOTICS, AND SYNBIOTICS

Prebiotics

Prebiotics are substrates to be fermented by flora (eg, nondigestible food ingredients, mainly plant fibers, that undigested reach the colon; food ingredients often referred to as colonic foods). Prebiotics have important functions in the body. They are essential to maintain mucosal growth and functions, to maintain water and electrolyte balance, to provide energy and nutrients for the host and for the flora, to enforce the body's resistance against invading pathogens, and to stimulate growth. Prebiotics are known to stimulate an increase in numbers and diversity of intestinal flora; relieve constipation and diarrhea; reduce serum triglycerides, serum cholesterol, and very-low-density lipoproteins; reduce the glycemic response to eating; improve water and electrolyte balance; and increase bioavailability and absorption of minerals such as calcium, magnesium, iron, and zinc.

Probiotics

Probiotics are live microorganisms supplied from the outside of the body, most commonly to the digestive tract. Flora and supplied probiotics have the ability to reduce or eliminate potentially pathogenic microorganisms from the body; reduce or eliminate the content of various toxins, mutagens, and carcinogens; promote apoptosis and disappearance of premalignant cells; release numerous nutrients, antioxidants, growth factors, coagulation factors, and other factors; modulate the innate and adaptive immune defense mechanisms; and stimulate gastrointestinal motility. Flora and supplied probiotics also are known to release numerous plant antioxidants; to synthesize some important vitamins, such as vitamin K, folic acid, niacin, thiamine, riboflavin, and B-complex vitamins [48]; to support mineral absorption [49]; and to reduce serum cholesterol [50]. A condition for clinical effect of supplied probiotics is that the supplied microbes remain viable during gastric and intestinal transit and that more than 10^7 colony-forming units/mL of viable bacteria reach the intestine [51]. Most probiotics available, such as those provided by dairy products or those sold in health stores, do not meet that condition. The ability to survive the acidity of the stomach and bile acid content of the small intestine seems to be limited to a few microorganisms. When a probiotic fruit drink (PRO VIVA; Probi, Lund, Sweden) containing no more than 10^7 of *L plantarum* 299V was tried in a controlled study of 129 patients undergoing abdominal surgery, no differences in bacterial translocation, gastric colonization with enteric organisms, or septic morbidity could be observed [52]; additionally, there were no differences in concentrations of plasma cells, IgA-positive cells, or IgM-positive cells in the lamina propria [53]. Great differences in ability of some LAB to survive and to influence cytokine production after passage through the stomach and small intestine were shown in a study in ileostomy patients [53]. Four different LAB species were compared: *L plantarum, L paracasei, L rhamnosus,* and *Bifidobacterium animalis*. Of originally orally administered 10^8 cells/mL of each LAB, after the passage through the stomach and small intestine only between 10^7 (*L plantarum*) and 10^2 (*L rhamnosus*) bacterial cells remained. Most of the strains tested showed after passage through the small intestine a significantly reduced or weak (especially *L rhamnosus*) ability to influence cytokine production (eg, the state of inflammation). If the study had included yogurt bacteria, even smaller survival and biologic effects could have been expected. The absence of clinical efficacy of yogurt bacteria when used in postoperative and severely sick patients has been documented in two controlled studies. A standard commercial product (TREVIS; Ch Hansen, Hoersholm, Denmark) containing *Lactobacillus acidophilus* LA5, *Bifidobacterium lactis* BP12, *Streptococcus thermophilus,* and *Lactobacillus bulgaricus* was mixed with 7.5 g of oligofructose and given to 45 critically ill patients [54] and 45 controls and to 72 elective abdominal surgery patients and 65 controls [55]. No clinical benefits were reported from either of the studies. The study in intensive care unit patients reported significant reductions in number of potentially pathogenic organisms in the stomach of the treated patients, but no influence on intestinal permeability or clinical benefits. The perioperative study reported no differences in bacterial translocation, gastric colonization, systemic inflammation, or septic complications. See also a commentary to the ICU study by Bengmark [56].

Only a few LAB can ferment semiresistant prebiotics, such as the oligofructans inulin and phleins. When the ability of 712 different LAB to ferment oligofructans was studied, only 16 of 712 were able to ferment the phleins, and only 8 of 712 were able to ferment the inulin-type fiber. Only four LAB species fermented these fibers: *L plantarum* (several strains), *L paracasei* subsp *paracasei, Lactobacillus brevis,* and *Pediococcus pentosaceus* [57]. Also the ability to control various pathogens is strain specific and often limited to a few strains. When the ability of 50 different LAB to control 23 different pathogenic CD organisms was tested, only 5 proved effective against all, 8 were antagonistic to some, and 27 were totally ineffective [58]. The five most effective strains were *L paracasei* subsp *paracasei* (two strains) and *L plantarum* (three strains). Information such as this is important for the choice of probiotics for clinical use.

Synbiotics

Synbiotics is the word coined for the combined treatment with specific bioactive LAB and specific prebiotics with ability to stimulate the growth of certain LAB and to provide definite health benefits by synergistic action. A condition for such effects is that the LAB used have a documented ability to metabolize simultaneously supplemented prebiotics, which is often not the case, especially when it comes to different oligosaccharides [57].

YOGURT AS CARRIER OF PROBIOTICS AND SYNBIOTICS

Cow's milk is not an ideal carrier of probiotics, especially for specific clinical use. In addition to its proposed role as risk factor for chronic diseases [9,10,59–62], concerns with cow's milk include the following:

1. Cow's milk is a poor carrier of bioactive fiber-fermenting probiotics because, in sharp contrast to breast milk, it contains no fibers or fiber-like molecules (only elephant milk contains as much as human milk). The complex fucosylated oligosaccharides in human milk, with structural similarities to immunomodulating cell surface glycoconjugates, protect breast-fed infants against infection and inflammation. These oligosaccharides most likely also serve as prebiotics, provide key nutrients to breast-fed infants, and stimulate growth of the nonpathogenic health-supporting gut microflora [63].
2. Cow's milk is known to release inflammatory mediators; induce inflammation; induce leakage of molecules, such as albumin and hyaluronan; increase intestinal permeability; and cause translocation and leaky gut [64–69].
3. Cow's milk is known to be rich in free polyunsaturated fatty acids. It was shown that presence of polyunsaturated fatty acids, even in lower concentrations than provided in fermented dairy products such as yogurt, cause *Lactobacillus* to lose their ability to adhere to mucous membranes and to grow, supporting that dairy products are not ideal as carriers of probiotics [70].

Supplementing with yogurt bacteria or similar bacteria yields small or no clinical benefits [54,55,71].

PROBIOTICS AND SYNBIOTICS IN CLINICAL STUDIES

The hope for identification of a "magic bug" capable of controlling all types of inflammation in all stages of disease seemingly has remained an illusion. The promising effects observed in experimental studies often have not been possible to repeat in patients, especially in patients with chronic diseases. Chronic diseases that develop spontaneously are more therapy-resistant than similar diseases when induced in animals. Induced diseases rarely remain chronic, and more dramatic effects of probiotics generally are obtained in acute conditions in animals and humans. Another possible explanation may be that animals usually receive much larger doses of probiotics in relation to mucosal surface and to body weight. No systematic dose-response studies have been published yet. The most successful results so far have been obtained with the use of cocktails of LAB, with or without simultaneous supplementation of prebiotics. These cocktails commonly are supplied in significantly larger doses, however, than is the case with the use of single-strain probiotics or single-strain/single-fiber synbiotics, which treatments most often are provided with daily doses of 1 to 10 billion LAB/d. The current trend is toward more complex compositions and toward use of much larger doses of LAB. Studies with the multistrain probiotic VSL#3 use doses between 1800 and 3600 billion LAB/d, and studies with the multistrain/multifiber Synbiotic 2000 use doses between 40 billion and 1200 billion LAB.

COMMERCIALLY AVAILABLE PREBIOTICS AND PROBIOTICS

Although some trials in the past have used LAB from health stores, dairy products, or a plethora of LAB available on the "spot market," most of the trials have been done with fewer than 10 different formulations:

Single-strain probiotics

Saccharomyces boulardi (Laboratories Biocodex, Montrouge, France) is commonly given in doses of 2 capsules containing 250 mg morning and evening, equivalent to approximately 10 billion live organisms/d. The nonpathogenic *E coli* serotype O6:K5:H1 (Mutaflor; Ardeypharm GmbH, Herdecke, Germany) is referred to as Nissle 1917 after early observations done by Nissle during the World War I. It is commonly given in doses of less than 10 billion LAB/d. *Lactobacillus GG* (LGG) (Valio, Helsinki, Finland) commonly is given in doses of 1 to 5 billion LAB/d. *L acidophilus* LA1 (LA1) (Nestle, Vevey, Switzerland) is commonly given in doses of less than 5 billion LAB/d (sometimes <1 billion LAB/d).

Multistrain probiotics

The probiotic cocktail VSL#3 (Sigma-Tau, Pomezia, Italy, and VSL Pharmaceuticals, Fort Lauderdale, Florida) is the only multistrain probiotic tried so far. It consists of four *Lactobacillus* strains (*L acidophilus, L casei, L delbruecki* subsp *bulgaricus,* and *L plantarum*), three *Bifidobacterium* strains (*B longum, B infantis, B breve*), and *S salivarius* ssp *thermophilus* (5×10^{11} cells/g). VSL#3 has been tried in several studies and yielded results that have attracted the interest of gastroenterologists and patients around the world. VSL#3 contains 300 billion live bacteria per gram. It is commonly given in high doses, usually 1800 billion LAB and more recently up to 3600 billion LAB/d.

Single-strain/single-fiber synbiotics

The composition of *L plantarum* 299 or 299V plus oat fiber (PRO VIVA; AB Probi, Lund, Sweden) was constructed after extensive studies of human *Lactobacillus* strains [72]. The composition contains 10 g of oat fiber and 10^9 (1 billion) LAB of *L plantarum*. A common dose is 1 or 2 billion LAB/d; occasionally 5 billion LAB/d has been tried. Most of the experience with this synbiotic stems from studies in critical care and in connection with extensive surgery.

Multistrain/multifiber synbiotics

Synbiotic 2000 (Medipharm, Kågeröd Sweden, and Des Moines, Iowa) consists of a mixture of four LAB, one from each of the four main genera of lactobacillus: 10^{10} of *P pentosaceus* 5-33:3, 10^{10} of *Leuconostoc mesenteroides* 32-77:1, 10^{10} of *L paracasei* subsp *paracasei* 19, and 10^{10} of *L plantarum* 2362 (eg, 40 billion LAB per dose, plus a mixture of four well-studied bioactive plant fibers: 2.5 g of β-glucan, 2.5 g of inulin, 2.5 g of pectin, and 2.5 g of resistant starch, total 10 g plant fibers. The composition was determined after extensive studies of more than 350 humans and more than 180 plant strains by Lund university microbiologists Ljungh and Wadström and their group [73,74]. They chose the LAB to be used in the composition based on the ability of the various LAB to produce bioactive proteins, transcribe nuclear factor-κB, produce proinflammatory and anti-inflammatory cytokines, produce antioxidants, and complement each other functionally. The four LAB individually function differently, but show synergistic effects when supplemented together. Newer supplements Synbiotic 2000 FORTE and Probiotic 2000 FORTE (no fiber added) are based on 10^{11} of each of the four LAB (eg, 400 billion LAB per dose, or if supplemented twice or three times daily 800 to 1200 billion LAB/d).

Total flora replacement

Total flora replacement (TFR) was introduced as a treatment alternative in severe *C difficile* infections. It is based on transfer of fecal flora from a healthy individual, often a close relative, to a severely sick patient, who is prepared for the supplementation by oral polyethylene glycol lavage and broad-spectrum antibiotic treatment. TFR has been used occasionally in severe *C difficile* cases, but also in severe constipation, IBS, and IBD. The patients usually receive, after a preceding washout, about 200 to 300 mL of fresh feces dissolved in an equal amount of saline solution. The process is repeated for about 5 to 7 days [75,76].

PROBIOTICS AND SYNBIOTICS IN GASTROINTESTINAL DISEASE

Infectious diarrhea

Diarrhea is one of the most common expressions of disease. In the developed world, mainly elderly and immunocompromised individuals are affected. In the developing world, diarrhea affects children with 6 to 12 episodes per year compared with about 2 episodes annually in children in the developed world. It is estimated that more than 3 million children die globally each year as a result of severe diarrhea. Reversing dehydration is the most effective treatment, but there is a strong need for complementary treatment. Prebiotics, probiotics, and synbiotics have the potential to be that complement because in addition to being relatively inexpensive and without serious side effects, they potentially can control infection and modulate motility. A meta-analysis based on 23 controlled studies totaling 1917 patients concluded that the risk of diarrhea was reduced by 3 days (relative risk 0.66) and the mean duration of diarrhea was 30.5 hours [77]. The effects are seemingly especially pronounced in rotavirus diarrhea, in which mean duration of diarrhea was reduced by 38.1 hours. The study identified great variations in results with the use of different probiotics. A combination of *L acidophilus* and *L bifidus* seemed to be the most effective. In contrast to most other regimens tried, *S thermophilus* and *L bulgaricus* seemed to have no effect on diarrhea, which is in line with observations in other conditions.

Antibiotic-associated diarrhea and *Clostridium difficile* colitis

Approximately one fourth of all antibiotic-associated diarrhea episodes involve *C difficile*. When 50 *C difficile* patients received 10 billion viable *S boulardi* organisms, recurrence was observed in 33 of 50 patients [78]. Prophylactic supplementation to antibiotic-treated patients of 20×10^6 colony-forming units/d of LGG was tried in a randomized trial involving 267 patients, but no difference was observed; diarrhea developed in 39 (29%) of LGG-treated patients and in 40 (30%) controls [79]. Twenty-nine patients with more than one verified episode of *C difficile*–associated diarrhea were supplemented with either 5 billion *L plantarum* 299V (LP299V) or placebo in addition to metronidazole [80]. No significant difference in outcome was observed; 4 of 11 LP299V-treated patients and 6 of 9 controls showed signs of recurrence. No study is yet reported with the use of more complex compositions, such as VSL#3 or Synbiotic 2000. TFR has been used in 84 patients: 36 patients with *C difficile*–associated diarrhea, 22 with *C difficile* colitis, and 26 with pseudomembranous colitis [75,76]; 72 of 84 patients (86%) showed immediate resolution of the problems. None of the patients as reported had signs of relapse during follow-up lasting 5 years. Cure is reported to have been achieved with a single-shot treatment in 33 of 36 patients (92%).

Crohn's disease

CD is a disease proven to be refractory to treatment [79]. With increasing knowledge about the inflammatory cascade, several possible treatment options have been suggested. So far antagonism of tumor necrosis factor (TNF)-α with the monoclonal antibody infliximab has proved to be the most successful. Future treatments are likely to be compared with infliximab as the gold standard or suggested as a supplement to infliximab treatment. Single-strain probiotics, such as *S boulardi* [81,82], *E coli* Nissle [83], and LGG [84,85], have been tried with no or minimal success. An abstract published in 2000 claimed an effect with VSL#3, but the expected complete article was never published [86]. Synbiotic 2000 has been tried in two controlled trials. After an initial treatment with infliximab, 63 patients were randomized to receive either Synbiotic 2000 or placebo daily [87]. Median time to relapse was 9.8 and 10.1 months. In the other study, 20 patients were supplemented with Synbiotic 2000, and 9 patients received placebo; no difference was observed in disease activity scores after 3 months (R. Eliakim, MD, PhD, personal communication, 2005). Bioecologic treatment has not been able contribute to improved outcome in this group of patients. One cannot exclude, however, that supply of significantly larger doses of probiotic and synbiotic compositions could prove effective.

Ulcerative colitis

Of single-strain probiotics, mainly *S boulardi* and *E coli* Nissle [88–90] have been tried, both with some, but far from satisfactory, improvements. In one *E coli* Nissle study involving 114 patients, 44 of 59 *E coli* Nissle–treated (75%) and 39 of 57 mesalamine-treated patients reached remission. Relapses during the study period occurred in 26 of 57 (67%) *E coli* Nissle–treated patients and 32 of 59 (73%) mesalamine-treated patients [89]. Subsequent reviewers have pointed out that the relapse rate in the mesalamine group was significantly higher than expected from the literature [91]. It also has been suggested that the study groups were heterogeneous with respect to severity of disease, and that the dose of mesalamine used in the study was lower than usually used. In the last study with *E coli* Nissle, 327 patients with quiescent UC were treated during 1 year with either *E coli* Nissle or mesalamine [90]. The relapse rate was 45% with *E coli* Nissle and 36% with mesalamine. A study using VSL#3 provided much more cutting-edge results [92]. For 1 year, 20 patients with UC in remission and intolerant to mesalamine treatment received daily morning and evening 3 g of pure VSL#3 bacteria, equivalent to, at that time, almost "astronomical amounts" of 1200 billion LAB. Four of the patients showed significant relapse after 3, 5, 5, and 7 months; 1 was lost to follow-up, the remaining 15 were still in remission after 12 months. Rectal instillations with Synbiotic 2000 reconstituted in saline were given to 10 patients during 2 weeks. One patient withdrew after 1 week; during the 3 weeks of observation, the remaining patients showed dramatic improvements in diarrhea scores, visible blood in stool, nocturnal diarrhea, urgency, and consistency of stool (Table 1) [93]. Two patients reported significant bloating and flatulence, but no other side effects were reported. Only 10 IBD patients, 9 UC patients, and 1 CD patient are reported as treated with TFR in the literature [75,76]. All TFR-treated IBD patients are reported to have had IBD for more than 5 years and to have been refractory to conventional treatments. The patients are claimed all to have experienced a complete reversal of disease and to have all anti-inflammatory treatments concluded after 6 weeks and, most important, to remain remission-free after periods of observation of 1 to 13 years. It also is claimed that on follow-up endoscopy and histology the mucosa had shown a normal

Table 1
Changes in urgency, episodes of diarrhea, nightly diarrhea, visible blood on stool, and consistency of stool during 2 weeks of repeat enemas with synbiotics and 3 weeks' study

	Day 1	Day 7	Day 14	Day 21
Urgency	1.9	1.2*	1*	1†
Episodes of diarrhea	2.4	1.3	0.9*	0.8*
Nightly diarrhea	0.5	0.1†	0‡	0‡
Visible blood	2.2	1.2†	0.8‡	0.8‡
Consistency of stool	1.1	0.9†	0.7‡	0.8‡

*$P < .05$.
†$P < .01$.
‡$P < .001$.

Data from Pathmakanthan S, Walsh M, Bengmark S, Willemse PJA, Bardhan KD. Efficacy and tolerability treating acute distal ulcerative colitis with synbiotic enemas: a pilot trial. Gut 2002;51(Suppl 1):A307.

appearance. This small but significant improvement with TFR is unprecedented. There is nothing to assume that even TFR would lead to more than a temporary colonization. Most likely the guest flora disappear within 2 to 4 weeks. With present knowledge, permanent neocolonization would not occur in adults. The most likely explanation seems at this stage to be that the guest flora has managed to eliminate an unknown, uncultivatable pathogen underlying the disease.

Pouchitis

LGG was tried in two studies of pouchitis. Twenty patients with a previous history of pouchitis and endoscopic signs of inflammation were randomized to either 1 billion LGG or placebo. No differences were observed in disease activity or total anaerobes and aerobes in mucosal biopsy specimens or in feces after 3 months of treatment [94]. Prophylactic supplementation of LGG immediately after surgery with construction of ileoanal anastomoses was tried in one study [95]. Comparison was with historical controls: 2 of 39 versus 8 of 78 developed single-episode pouchitis, 1 versus 12 had recurrent episodes of pouchitis, and 0 versus 7 had chronic pouchitis. Cutting-edge improvements were reported from a controlled study with supplementation of 1200 billion VSL#3 for 9 months. Only 3 of 20 VSL#3-treated patients in contrast to 20 of 20 control patients had remission of pouchitis [96]. Similar results were obtained in a second study in collaboration with British gastroenterologists; 17 of 20 (85%) VSL#3-treated and 1 of 16 controls remained in remission after 1 year of treatment [97]. In a third study, supplementation of VSL#3 was started immediately after the surgical construction of ileoanal anastomosis with the hope to prevent pouchitis from developing. Only 2 of 20 (10%) patients in the VSL#3-treated group compared with 8 of 20 (40%) in the control group ($P < .01$) developed pouchitis [98].

Irritable bowel syndrome

IBS is likely the most common gastrointestinal disorder in the Western world. It is characterized by altered bowel habits, dysmotility, and abdominal pain or discomfort. The pathogenesis is obscure, but a study suggests that most patients (34 of 44) have a significant ($P < .001$) infiltration with mast cells in the mucosa compared with controls [99]. Especially the number of mast cells located close to (within 5 μ) nerve fibers are significantly increased in IBS patients and significantly related to degree of abdominal pain and discomfort. Probiotics have been shown to modulate the enteroendocrine cell population in the rat intestine [100]. Several attempts to affect the disease with probiotic supplementation are reported in the literature. Twenty-four patients were randomized to receive either 5 billion LGG or placebo twice daily during 10 weeks, but no difference between the two groups was observed in bloating, pain scores, or bowel movements [101]. In a well-controlled study, 78 patients were supplemented with 10 billion of either *L salivarius* UCC4331 or *B infantis* 35624 or placebo [102]. Two different composite scores based on the three cardinal symptoms–pain/discomfort, bloating/distention and bowel movement difficulty–were calculated, and both showed statistically significant improvement with probiotic supply, the most pronounced effect seen with supply of *B infantis*. The abnormal ratio of the cytokines interleukin-10 to interleukin-12 was reported normalized in the *B infantis*–treated group. Dual-strain probiotics were tried in a study in which 50 patients were randomized to receive a combination of 5 billion of *L plantarum* Lp01 and 5 billion of *B breve* Br0 during 1 month. Reductions in pain and overall symptom scores were reported as 50% and 25%; no statistical analysis was regarded possible [103].

VSL#3 in a dose of 450 billion LAB/d was tried in a controlled study in 25 patients at the Mayo Clinic [104]. After 8 weeks, no differences were observed between VSL#3-treated and controls in mean gastrointestinal transit time, abdominal pain, gas, and urgency. The single-strain/single-fiber synbiotic based on *L plantarum* 299V and oat fiber was tried in three studies [105–107]. Twenty of 20 patients in the *L plantarum* 299V–treated group and 11 of 20 in the placebo group reported resolution of their abdominal pain [105]. A similarly designed study in 2 × 30 patients found the most modest influence of *L plantarum* 299V treatment–reduction in flatulence but no influence on bloating [106]. The third study reported no effect at all of *L plantarum* 299V treatment [107]. Total flora replacement also has been tried in a few patients, but so far the experience is at best anecdotal.

Helicobacter pylori infections

In sharp contrast to other microbes, a few *Lactobacillus* strains have the ability to tolerate low pH and survive and grow in the otherwise hostile-to-microbes environment of the stomach [108]. This ability could offer unique opportunities to prevent overgrowth of *H pylori,* the main cause of chronic gastritis and peptic ulcers and an important risk factor for gastric malignancies. A study compared 17 different LAB and their ability to inhibit growth of 10 different *H pylori* strains [109]. All strains inhibited the growth at low pH, but at a pH of 6.0, all strains except *L acidophilus* CRL639 lost that ability. This observation speaks against simultaneous use of H_2 blocking agents. A total of 120 patients were randomized to receive *L acidophilus* or placebo as supplement to a 7-day course of triple-agent therapy (rabeprazole, clarithromycin, amoxicillin) [110]. The eradication rates were 52 of 59 (88%) and 42 of 58 (72%; $P = .03$). When the study was repeated with LGG, however, no such triple therapy–potentiating effect could be observed [111,112]. Daily consumption of 4 × 50 mL of supernatant from a whey-based culture with *L acidophilus* is reported to reduce significantly breath test results [113], and to decrease density of *H pylori* in the stomach [114]. Subsequent studies with *L acidophilus* [69] and a variety of other *Lactobacillus* and *Bifidobacterium* species [115,116] seem, however, to show universal suppression of *H pylori* by various LAB, which seemingly is independent of strains used [116].

Chronic liver disease

Endotoxin-induced activation of macrophages is assumed to be responsible for increased levels of circulating TNF-α and soluble TNF receptor observed in cirrhotics and the expression of toll-like receptors TLR2 and TLR4. Circulating endotoxin, TNF-α and soluble TNF receptor levels, peripheral blood mononuclear cell expression of TLR2 and TLR4, and in vitro TNF-α production by peripheral blood mononuclear cells when stimulated with endotoxin or *S aureus* enterotoxin B were measured in 36 cirrhotic patients supplemented Synbiotic 2000 and 32 controls [117]. Supplementation of synbiotics resulted in significant up-regulation of peripheral blood mononuclear cell expression of TLR2.

In a subsequent study, 55 patients with minimal hepatic encephalopathy were randomized to receive Synbiotic 2000 ($n = 20$), only the fiber in Synbiotic 2000 ($n = 20$), or placebo ($n = 15$) for 30 days [47]. Patients with cirrhosis and minimal hepatic encephalopathy were found to have substantial derangements in the gut microecology, with significant fecal overgrowth of potentially pathogenic *E coli* and staphylococcal species. Synbiotic treatment significantly increased fecal content of non–urease-producing *Lactobacillus* species and significantly reduced endotoxemia and potentially pathogenic flora. The observed alterations in gut flora were accompanied by a significant reduction in blood ammonia levels and reversal of minimal hepatic encephalopathy in half of the patients; the Child-Turcotte-Pugh functional class improved in nearly 50% of cases. Treatment with only fermentable fiber was beneficial in a substantial proportion of patients. Interventions aimed at reducing intestinal levels of endotoxin-containing, gram-negative bacteria have been suggested also to improve systemic hemodynamic disturbance in cirrhosis, but effects on hepatic blood flow were not reported. A study by the same group performed in 15 cirrhotic patients showed a significant reduction (median 17.5%, range 1.4–65%) in indocyanine green retention at 15 minutes (ICG_{R15}) in the cirrhotic patients after 7 days of supplementation with Synbiotic 2000 ($P = .003$) [118].

Orthotopic liver transplantation

Two prospective randomized trials with supplementation of synbiotics also have been performed. In the first study, the single-strain/single-fiber synbiotic preparation based on 1 billion *L plantarum* 299 and 10 g oat fiber (L299) was compared with 1 billion heat-killed L299 and 10 g oat fiber (H299) and with selective digestive tract decontamination (SDD) [119]. The study comprised 95 patients divided into three groups: (1) SDD four times daily for 6 weeks ($n = 32$), (2) L299 ($n = 31$) during 12 postoperative days, and (3) H299 ($n = 32$) during 12 postoperative days. The same enteral nutrition was supplied to all patients. There were no deaths. Signs of infections occurred in 48% (15 of 32) of SDD patients, in 34% (11 of 32) of H299 patients, and in 13% (4 of 31) of L299 patients ($P = .017$). The numbers of postoperative infections were SDD, 23; H299, 17; and L299, 4. The numbers of patients requiring hemodialysis were SDD, 8; H299, 4; and L299, 2. In a subsequent double-blind randomized study, 33 patients were supplemented with the multistrain/multifiber Synbiotic 2000, and another 33 patients received only the four fibers in the synbiotic composition [120]. The treatment started on the day before surgery and continued until day 14 after surgery. Only one patient in the Synbiotic 2000–treated group (3%) showed signs of infection (urinary infection) during the first month compared with 17 of 33 (51%) patients supplemented with only the four fibers. The infecting bacterium with Synbiotic 2000 was *Enterococcus faecalis* in 1 patient compared with 11 in the fiber-only group. The group supplied fiber only also had *E coli* infection in three patients, *Enterobacter cloacae* infection in two patients, *Pseudomonas aeruginosa* infection in two patients, and *S aureus* infection in one patient. The use of antibiotics was also significantly shorter in the Synbiotic 2000–treated group.

Perioperative and critical care

In a prospective randomized study, the effect of the single-strain/single-fiber synbiotic based on *L plantarum* 299 in a dose of 1 billion LAB/d and 10 g of oat fiber (L299) was compared with a similar amount of heat-killed *L plantarum* 299 and oat fiber (H299) and parenteral nutrition (PN) in 3 × 30 patients undergoing abdominal operations, including liver resection, pancreas resection, gastric resection, colon resection, and intestinal bypass [121]. The L299 and H299 groups had significantly fewer infections (3 of 30 patients in each group, 10%) compared with the PN group (9 of 30 patients, 30%; $P > .001$). An even larger difference was observed when the subgroup of gastric and pancreatic surgery patients was analyzed separately: None of the L299 patients, one of eight H299 patients (12%), and three of six (50%) PN patients had infections. A similar study was done in patients undergoing abdominal cancer operations. Postoperative infections were observed in 1 of 15 patients (6.7%) when the multistrain/multifiber Synbiotic 2000 was supplemented in a dose of 40 billion LAB/d, in 3 of 15 patients (20%) supplemented with only the fibers in Synbiotic 2000, and in 7 of 15 patients (47%) supplemented with standard enteral nutrition (Han Chun Mao, personal communication). Significant improvements in prealbumin, C-reactive protein, serum cholesterol, serum endotoxin, and white cell blood count also were observed. A more recent, not-yet-published, study in acute extensive trauma patients reported a dramatic decrease in number of chest infections with supplementation of 40 billion LAB-containing Synbiotic 2000 (1 of 14 patients, 7%) compared with only fibers (11 of 28 patients, 39%), peptide (10 of 21 patients, 48%), or glutamine (12 of 37 patients, 32%). Equally the total number of infections were decreased: Synbiotic 2000, 2 of 14 patients (14%); only fiber, 16 of 28 patients (57%); peptide, 11 of 21 patients (52%); and glutamine, 19 of 37 patients (51%) (L. Kompan, MD, PhD, personal communication, 2005).

SUMMARY

Prebiotic, probiotic, and synbiotic treatment is still in its infancy. Although remarkable effects have been observed, extensive studies are necessary to understand the many mechanisms behind the observed effects. Compared with acute conditions, chronic diseases seem much more resistant to attempts to affect the course by modification of microbiota. Many of the significant effects observed in experimental animals with induced chronic diseases have not been repeated in patients with chronic diseases, especially patients with IBD. Induced chronic diseases are more acute than chronic, which might explain the differences in sensitivity to modulation of microbiota. Also, there is a tendency to use much higher doses in animals than has hitherto been the case in humans. The experience with total flora replacement cannot be ignored; dramatic effects sometimes have been observed in the patients receiving transfer of extensive amounts of normal human flora and fibers. Dramatic effects also have been observed with the introduction of VSL#3, which contains more strains and is provided in much larger amounts–sometimes several times more than has hitherto been the case. It is disturbing, however, that the producers have provided no information about the criteria for selection of the strains in the composition. With the lack of such information, one has to assume that the selection has been at random. It is hoped that such information will be made available. As pointed out by Shahanan [122]: "as with all medications, it is preferable that the properties and behavior of an individual component of probiotic cocktails be fully determined in vitro, with synergistic and antagonistic activities identified, before they are adopted for widespread routine use." Most likely several of the strains in the composition, such as the yogurt bacteria, do not add to the efficacy of the probiotic cocktail. Nevertheless, VSL#3 has provided the direction for future avenues in the development of more efficient tools for modification of microbiota. Most likely, future research will focus on attempts to use synergistic effects by combining several strains and attempts to improve efficacy by the use of much larger doses than previously used. Most likely doses of 5 trillion LAB/d will be tried. In an excellent review, Sartor [123] concluded that "the interesting approach of combining probiotic and prebiotic agents (synbiotics) has considerable appeal." The extensive experience in acute diseases and the limited experience in chronic liver disease with synbiotic treatment seem to support such an assumption. As stated by Sartor [123] in his comprehensive review, "current data for therapeutic efficacy do not withstand rigorous scrutiny or fulfill current evidence based rationale for using antibiotics, probiotics and prebiotics in the treatment of IBD," and "clinical trials have consistently been underpowered to show equivalency or superiority, many have design flaws that preclude definite results, or use outcomes such as disease activity index, that do not conform with widely accepted criteria for disease response or remission" Too often "enthusiasm outstrips scientific support for these therapeutic approaches." It is crucial that studies be performed by different and independent groups on all continents with their different living conditions.

Recommended standards for clinical trials and specifically for IBD are available [124,125]. It seems that none of the IBD-related studies have attempted to meet these standards. Almost 100 years have passed since Metchnikoff [126] suggested health benefits from external health-promoting bacteria and almost 50 years since Eiseman et al [127] reported successful treatment with fecal enemas in antibiotic-associated pseudomembranous enterocolitis. Clinicians since have not administered the inheritance well.

References

[1] Bengmark S. Acute and "chronic" phase response—a mother of disease. Clin Nutr 2004;23:1256–66.

[2] Lin BF, Huang CC, Chiang BL, Jeng SJ. Dietary fat influences Ia antigen expression, cytokines and prostaglandin E_2 production in immune cells in autoimmune-prone NZBxNZW F1 mice. Br J Nutr 1996;75:711–22.

[3] Watanabe S, Onozaki K, Yamamoto S, Okuyama H. Regulation by dietary essential fatty acid balance of tumor necrosis factor production in mouse macrophages. J Leukoc Biol 1993;53:151–6.

[4] Roszkowski K, Ko KL, Beuth J, et al. Intestinal microflora of BALB/c-mice and function of local immune cells. Zeitschrift für Bakteriologie und Hygien 1988;270:270–9.

[5] Pulverer G, Beuth J, Roszkowski W, et al. Bacteria of human physiological microflora liberate immunomodulating peptides. Zentralbl Bakteriol 1990;272:467–76.

[6] Larsen TB, Nielsen JN, Fredholm L, et al. Platelets and anticoagulant capacity in patients with inflammatory bowel disease. Pathophysiol Haemost Thromb 2002;32:92–6.

[7] Bjerregaard LT, Nederby NJ, Fredholm L, et al. Hyperhomocysteinaemia, coagulation pathway activation and thrombophilia in patients with inflammatory bowel disease. Scand J Gastroenterol 2002;37:62–7.

[8] Capristo E, Mingrone G, Addolorato G, Greco AV, Gasbarrini G. Glucose metabolism and insulin sensitivity in inactive inflammatory bowel disease. Aliment Pharmacol Ther 1999;13:209–17.

[9] Hunter JO, et al. Dietary studies. In: Gibson PR, Jewell DP, editors. Topics in gastroenterology. Oxford (UK): Blackwell Science; 1985. p. 305–13.

[10] Holmes MD, Pollak MN, Willett WC, Hankinson SE. Dietary correlates of plasma insulin-like growth factor I and insulinlike growth factor binding protein 3 concentrations. Cancer Epidemiol Biomarkers Prev 2002;11:852–61.

[11] Braly J, Hoggan MA. Dangerous grains. New York: Avery: Penguin Putnam Inc; 2002.

[12] Fine KD, Do K, Schulte K, et al. High prevalence pf celiac sprue-like HLA-DQ genes and enteropathy in patients with microscopic colitis syndrome. Am J Gastroenterol 2000;95:1974–82.

[13] Di Cagno R, De Angelis M, Auricchio S, Greco L, Clarke C, De Vincenzi M. Sourdough bread made from wheat and nontoxic flours and started with selected lactobacilli is tolerated in celiac sprue patients. Appl Environ Microbiol 2004;70:1088–96.

[14] Bengmark S. Prospect for a new and rediscovered form of therapy: probiotic and phage. In: Andrew PW, Oystron P, Smith GL, Stewart-Tull DE, editors. Fighting infection in the 21st century. Oxford (UK): Blackwell Scientific; 2000. p. 97–132.
[15] Gustafsson B. The future of germ-free research. New York: Alan R: Liss; 1985.
[16] Hooper LV, Midtvedt T, Gordon JI. How host-microbial interactions shape the nutrient environment of the mammalian intestine. Annu Rev Nutr 2002;22:283.
[17] Alverdy JC, Laughlin RS, Wu R. Influence of the critically ill state on host pathogen interactions within the intestine: gut derived sepsis redefined. Crit Care Med 2003;31:598–607.
[18] Leveau P, Wang X, Soltesz V, Ihse I, Andersson R. Alterations in intestinal permeability and microflora in experimental acute pancreatitis. Int J Pancreat 1996;20:119–25.
[19] Andersson R, Wang X, Ihse I, Leveau P, Ihse I. The influence of abdominal sepsis on acute pancreatitis in rats: a study on mortality, permeability, arterial blood pressure and intestinal blood flow. Pancreas 1995;11:365–73.
[20] Knight DJW, Ala'Aldeen D, Bengmark S, Girling KJ. The effect of synbiotics on gastrointestinal flora in the critically ill [abstract]. Br J Anaesth Abstract 2004;92:307–8.
[21] Lencner AA, Lencner CP, Mikelsaar ME, et al. Die quantitative Zusammensetzung der Lactoflora des Verdauungstrakts vor und nach kosmischen Flügen unterschiedlicher Dauer. Nahrung 1984;28:607–13.
[22] Sepp E, Julge K, Vasur M, Naaber P, Bjorksten B, Mikelsaar M. Intestinal microflora of Estonian and Swedish infants. Acta Pediatr 1997;86:956–61.
[23] Adlerberth I, Carlsson B, deMan P, et al. Intestinal colonization with *Enterobacteriaceae* in Parkistani and Swedish hospital-delivered infants. Acta Pediatr Scand 1991;80: 602–10.
[24] Finegold SM, Sutter VL, Mathisen GE. Normal indigenous intestinal flora. In: Hentges DJ, editor. Human intestinal microflora in health and disease. London: Academic Press; 1983. p. 3–31.
[25] Ahrné S, Nobaek S, Jeppsson B, Adlerberth I, Wold AE, Molin G. The normal *Lactobacillus* flora in healthy human rectal and oral mucosa. J Appl Microbiol 1998;85:88–94.
[26] Lorber B. Are all diseases infectious? Ann Intern Med 1996;125:844–51.
[27] Fabia R, Ar'Rajab A, Johansson ML, et al. Impairment of bacterial flora in human ulcerative colitis and in experimental colitis in the rat. Digestion 1993;54:248–55.
[28] Ott SJ, Wenderoth DF, Hampe J, et al. Reduction in diversity of the colonic mucosa associated bacterial microflora in patients with active inflammatory bowel disease. Gut 2004;53:685–93.
[29] Swidsinski A, Ladhoff A, Pernthaler A, et al. Mucosal flora in inflammatory bowel disease. Gastroenterology 2002;122:44–54.
[30] Darfeuille-Michaud A, Boudeau J, Bulois P, et al. High prevalence of adherent-invasive *Escherichia coli* associated with ileal mucosa in Crohn's disease. Gastroenterology 2004;127:412–21.
[31] Grisham MD. Oxidants and free radicals in inflammatory bowel disease. Lancet 1994;344:859–61.
[32] Gross V, Arndt H, Andus T, Palitzsch KD, Scholmerich J. Free radicals in inflammatory bowel diseases: pathophysiology and therapeutic implications. Hepatogastroenterology 1994;41:320–7.
[33] Sido B, Hack V, Hochlehnert A, Lipps H, Herfarth C, Dröge W. Impairment of intestinal glutathione synthesis in patients with inflammatory bowel disease. Gut 1998;42: 485–92.
[34] Miralles-Barrachina O, Savoye G, Belmonte-Zalar L, et al. Low levels of glutathione in endoscopic biopsies in patients with Crohn's colitis: the role of malnutrition. Clin Nutr 1999;18:313–7.
[35] Park EJ, Jeon CH, Ko G, Kim J, Sohn DH. Protective effects of curcumin in rat liver injury induced by carbon tetrachloride. J Pharm Pharmacol 2000;52:437–40.

[36] Nanji AA, Jokelainen K, Tipoe GL, Rahemtulla A, Thomas P, Dannenberg AJ. Curcumin prevents alcohol-induced liver disease in rats by inhibiting the expression of KF-κB-dependent genes. Am J Physiol Gastroenterol Liver Physiol 2003;284:G321–7.
[37] Sahl B, Assi V, Templeman V, et al. Curcumin attenuates DBN-induced colitis. Am J Physiol Gastroenterol Liver Physiol 2003;285:G235–43.
[38] Ukil A, Maity S, Karmakar S, Datta N, Vedasiromoni JR, Das PK. Curcumin, the major component of food flavour turmeric, reduces mucosal injury in trinitrobenzene sulphonic acid-induced colitis. B J Pharmacol 2003;139:209–18.
[39] Fernadez-Banares F, Hinojosa J, Sanchez-Lombrana JL, et al. Randomized clinical trial of *Plantago ovata* seeds (dietary fiber) as compared with mesalamine in maintaining remission in ulcerative colitis. Am J Gastroenterol 1999;94:427–33.
[40] Kanauchi O, Iwanaga T, Mitsuyama K. Germinated barley foodstuff feeding: a novel neutraceutical strategy for ulcerative colitis. Digestion 2001;63(Suppl 1):S60–7.
[41] Wischmeyer PE, Riehm J, Singleton KD, et al. Glutamine attenuates tumor necrosis factor-α release and enhances heat shock protein 72 in human peripheral blood mononuclear cells. Nutrition 2003;19:1–6.
[42] Hallert C, Björck I, Nyman M, Pousette A, Granno C, Svensson H. Increasing fecal butyrate in ulcerative colitis patients by diet: controlled pilot study. Inflamm Bowel Dis 2003;9: 116–21.
[43] Welters CFM, Heineman E, Thunnissen FBJM, van den Bogaard AEJM, Soeters PB, Baeten CGMI. Effect of dietary inulin supplementation on inflammation of pouch mucosa in patients with an ileal pouch-anal anastomosis. Dis Colon Rectum 2002;45:621–7.
[44] Hopkins MJ, Macfarlane GT. Nondigestible oligosaccharides enhance bacterial colonization resistance against *Clostridium difficile* in vitro. Appl Environ Microbiol 2003;69:1920–7.
[45] Dasarathy S. Role of gut bacteria in the therapy of hepatic encephalopathy with lactulose and antibiotics. Indian J Gastroenterol 2003;22(Suppl 2):S50–3.
[46] Rabbani GH, Teka T, Saha SK, et al. Clinical studies in persistent diarrhea: dietary management with green banana or pectin in Bangladesh children. Gastroenterology 2001;121:554–60.
[47] Liu Q, Duan ZP, Ha da K, Bengmark S, Kurtovic J, Riordan SM. Synbiotic modulation of gut flora: effect on minimal hepatic encephalopathy in patients with cirrhosis. Hepatology 2004;39:1441–9.
[48] Tamine AY, Marshall VME, Robinson RK. Microbiological and technical aspects of milks fermented by bifidobacteria. J Dairy Res 1995;62:151–87.
[49] Deguchi Y, Morishita T, Mutai M. Comparative studies on synthesis of water-soluble vitamins among human species of bifidobacteria. J Agric Biol Chem 1985;49:13–9.
[50] Tamii Y, Yoshimitsu N, Watanabe Y, Kuwabara Y, Nagai S. Effects of milk fermented by culturing with various lactic acid bacteria and a yeast on serum cholesterol level in rats. J Ferment Bioeng 1996;81:181–2.
[51] Gomes AMP, Malcata FK. *Bifidobacterium* spp. and *Lactobacillus acidophilus*: biological, biochemical, technological and therapeutical properties relevant for use as probiotics. Trends Food Sci Tehnol 1999;10:139–57.
[52] McNaught CE, Woodcock NP, MacFie J, Mitchell CJ. A prospective randomized study of the probiotic *Lactobacillus plantarum* 299V on indices of gut barrier function in elective surgical patients. Gut 2002;51:827–31.
[53] Miettinen M, Alander M, von Wright A, et al. The survival of and cytokine induction by lactic acid bacteria after passage through a gastrointestinal model. Microb Ecol Health Dis 1998;10:41–7.
[54] Jain PK, McNaught CE, Anderson ADG, MacFie J, Mitchell CJ. Influence of synbiotic containing *Lactobacillus acidophilus LA5, Bifidobacterium lactis BP 12, Streptococcus thermophilus, Lactobacillus bulgaricus* and oligofructose on gut barrier function and sepsis in critically ill patients: a randomized controlled trial. Clin Nutr 2004;23:467–75.

[55] Woodcock NP, McNaught CE, Morgan DR, Gregg KL, MacFie J. An investigation into the effect of a probiotic on gut immune function in surgical patients. Clin Nutr 2004;23: 1069–73.
[56] Bengmark S. Synbiotics to strengthen gut barrier function and reduce morbidity in critically ill patients. Clin Nutr 2004;23:441–5.
[57] Müller M, Lier D. Fermentation of fructans by epiphytic lactic acid bacteria. J Appl Bacteriol 1994;76:406–11.
[58] Naaber P, Smidt I, Stsepetova J, Brilene T, Annuk H, Mikelsaar M. Inhibition of *Clostridium difficile* strains by intestinal *Lactobacillus* species. Med Microbiol 2004;53:551–4.
[59] Moss Moss M, Freed DL. Survival trends, coronary event rates, and the MONICA project: monitoring trends and determinants in cardiovascular disease. Lancet 1999;354:862–4.
[60] Qin L-Q, Wang P-Y, Kaneko K, Sato A. Estrogen: one of the risk factors in milk for prostate cancer. Med Hypoth 2004;62:133–42.
[61] Outwater JL, Nicholson A, Barnard N. Dairy products and breast cancer: the IGF-1, estrogen and bGH hypothesis. Med Hypoth 1997;48:453–61.
[62] Solomons NW. Nature's perfect food revisited: recent insights on milk consumption and chronic disease risk. Nutr Rev 2002;60:180–2.
[63] Gnoth MJ, Kunz C, Kinne-Saffran E, Rudloff S. Human milk oligosacharides are minimally digested in vitro. J Nutr 2000;130:3014–20.
[64] Kalliomaki M, Salminen S, Arvilommi H, et al. Probiotics in primary prevention of atopic disease: a randomized placebo-controlled trial. Lancet 2001;357:1076–9.
[65] Jalonen T. Identical intestinal permeability changes in children with different clinical manifestations of cow's milk allergy. J Allergy Clin Immunol 1991;88:737–42.
[66] Isolauri E, Majamaa H, Arvola T, Rantala I, Virtanen E, Arvilommi H. *Lactobacillus casei* strain GG reverses increased intestinal permeability induced by cow milk in suckling rats. Gastroenterology 1993;105:1643–50.
[67] Bengtsson U, Knutson TW, Knutson L, Dannaeus A, Hallgren R, Ahlstedt S. Increased levels of hyaluronan and albumin after intestinal challenge in adult patients with cow's milk intolerance. Clin Exp Allergy 1996;26:96–103.
[68] Bengtsson U, Knutson TW, Knutson L, Dannaeus A, Hallgren R, Ahlstedt S. Eosinophil cationic protein and histamine after intestinal challenge in patients with cow's milk intolerance. Allerg Clin Immunol 1997;100:216–21.
[69] Ehn BM, Ekstrand B, Bengtsson U, Ahlstedt S. Modification of IgE binding during heat processing of the cow's milk allergen. J Agric Food Chem 2004;52:1398–403.
[70] Kankanpää PE, Salminen SJ, Isolauri E, Lee YK. The influence of polyunsaturated fatty acids on probiotic growth and adhesion. FEMS Microbiol Lett 2001;194:149–53.
[71] Cruchet S, Obregon MC, Salazar G, Diaz E, Gotteland M. Effect of the ingestion of a dietary product containing *Lactobacillus johnsonii* La1 on *Helicobacter pylori* colonization in children. Nutrition 2003;19:716–21.
[72] Johansson ML, Molin G, Jeppsson B, Nobaek S, Ahrne S, Bengmark S. Administration of different lactobacillus strains in fermented oatmeal soup: in vivo colonization of human intestinal mucosa and effect on the indigenous flora. Appl Environ Microbiol 1993;59: 15–20.
[73] Kruszewska K, Lan J, Lorca G, Yanagisawa N, Marklinder I, Ljungh Å. Selection of lactic acid bacteria as probiotic strains by in vitro tests. Proceedings of the XVI International Congress on Microbial Ecology and Disease, Noordwijkerhout, The Netherlands, October 2001. Microecology and Therapy 2002;29:37–51.
[74] Ljungh Å, Lan J-G, Yamagisawa N. Isolation, selection and characteristics of *Lactobacillus paracasei* ssp *paracasei* isolate F19. Microb Ecol Health Dis 2002;3(Suppl):4–6.
[75] Borodo TJ, Warren EF, Leis S, Surace R, Ashman O. Treatment of ulcerative colitis using fecal bacteriotherapy. J Clin Gastroenterol 2003;37:42–7.
[76] Borody TJ, Warren EF, Leis SM, Surace R, Ashman O, Siarakas S. Bacteriotherapy using fecal flora: toying with human motions. J Clin Gastroenterol 2004;38:475–83.

[77] Allan SJ, Okoko B, Martinez E, Gregorio G, Dans LF. Probiotics for treating infectious diarrhoea. The Cochrane Database of Systematic Reviews 2003, issue 4, Art No CD003048 pub2, DOI:10.1002/14651858.CD003048.pub2.
[78] Elmer GW, McFarland LV, Surawics CM, Danko L, Surawics Danko L, Greenberg RN. Behavior of *Sacharomyces boulardii* in recurrent *Clostridium difficile* disease patients. Aliment Pharmacol Ther 1999;13:1663–8.
[79] Thomas MR, Litin SC, Osmon DR, Corr AP, Weaver AL, Lohse CM. Lack of effect of lactobacillus GG on antibiotic-associated diarrhea: a randomized placebo-controlled trial. Mayo Clin Proc 2001;76:883–9.
[80] Wullt M, Johansson-Hagslätt ML, Odenholt I. *Lactobacillus plantarum* 299v for the treatment of recurrent *Clostridium difficile*-associated diarrhoea: a double blind, placebo-controlled trial. Scand J Infect Dis 2003;35:365–7.
[81] Plein K, Hotz J. Therapeutic effects of *Saccharomyces boulardii* on mild residual symptoms in a stable phase of Crohn's disease with special respect to chronic diarrhea—a pilot study. Z Gastroenterol 1993;31:129–34.
[82] Guslandi M, Mezzi G, Sorghi M, Testoni PA. *Saccharomyces boulardi* in maintenance treatment of Crohn's disease. Dig Dis Sci 2000;45:1462–4.
[83] Malchow HA. Crohn's disease and *Escherichia coli*: a new approach in therapy to maintain remission in colonic Crohn's disease? J Clin Gastroenterol 1997;25:653–8.
[84] Prantera C, Scribano ML, Falasco G, Andreoli A, Luzi C. Ineffectiveness of probiotics in preventing recurrence after curative resection of Crohn's disease: a randomised controlled trial with *Lactobacillus* GG. Gut 2002;51:405–9.
[85] Schultz M, Timmer A, Herfarth HH, Sartor RB, Vanderhoof JA, Rath HC. *Lactobacillus* GG in inducing and maintaining remission of Crohn's disease. BMC Gastroenterol 2004;4:1–4.
[86] Campieri M, Rizzello F, Venturi A, et al. Combination of antibiotic and probiotic treatment is efficacious in prophylaxis of post-operative recurrence of Crohn's disease: a randomized controlled study vs mesalamine [abstract]. Gastroenterology 2000;118:A4279.
[87] Rutgeerts P, D'Haens G, Baert F, et al. Randomized placebo controlled trial of pro- and prebiotics (synbiotics cocktail) for maintenance of infliximab induced remission of luminal Crohn's disease (CD). Digestive Week USA, May 2004. Gastroenterology 2004;126:A467.
[88] Kruiz W, Schutz E, Fric P, Fixa B, Judmaier G, Stolte M. Double-blind comparison of an oral *Escherichia coli* preparation and meselazine in maintaining remission of ulcerative colitis. Aliment Pharmacol Ther 1997;11:853–8.
[89] Rembacken BJ, Snelling AM, Hawkey PM, Chalmers DM, Axon AT. Non-pathogenic *Escherichia coli* versus mesalazine for the treatment of ulcerative colitis: a randomised trial. Lancet 1999;354:635–9.
[90] Kruiz W, Fric P, Stolte M. Maintenance of remission in ulcerative colitis is equally effective with *Escherichia coli* Nissle 1997 and with standard mesalamine. Gastroenterology 2001;120:A139.
[91] Faubion WA, Sandborn WJ. Probiotic therapy with *E coli* for ulcerative colitis: take the good with the bad. Gastroenterology 2000;118:650–5.
[92] Venturi A, Gionchetti P, Rizzello F, et al. Impact on the composition of the faecal flora by a new probiotic preparation: preliminary data on maintenance treatment of patients with ulcerative colitis. Aliment Pharmacol Ther 1999;13:1103–8.
[93] Pathmakanthan S, Walsh M, Bengmark S, Willemse PJA, Bardhan KD. Efficacy and tolerability treating acute distal ulcerative colitis with synbiotic enemas: a pilot trial. Gut 2002;51(Suppl 1):A307.
[94] Kuisma J, Mentula S, Jarvinen H, Kahri A, Saxelin M, Farkkila M. Effect of *Lactobacillus rhamnosus* GG on ileal pouch inflammation and microbial flora. Aliment Pharmacol Ther 2001;17:509–15.

[95] Gosselink MP, Schouten WR, van Lieshout LMC, Hop WCJ, Laman JD, Ruseler-van Embden JGH. Delay of first onset of pouchitis by oral intake of the probiotic strain *Lactobacillus rhamnosus* GG. Dis Colon Rectum 2004;47:876–84.
[96] Gionchetti P, Rizzello F, Venturi A, et al. Oral bacteriotherapy as maintenance treatment in patients with chronic pouchitis: a double-blind, placebo-controlled trial. Gastroenterology 2000;119:305–9.
[97] Mimura T, Helwig U, Paggioli G, et al. One daily high dose probiotic therapy (VSL#3) for maintaining remission in recurrent or refractory pouchitis. Gut 2004;53:108–14.
[98] Gionchett P, Rizzello F, Helwig U, et al. Prophylaxis of pouchitis onset with probiotic therapy: a double-bland, placebo-controlled trial. Gastroenterology 2003;124:1202–9.
[99] Stanghellini BG, De Giorgio R, Cremon C, et al. Activated mast cells in proximity to colonic nerves correlate with abdominal pain in irritable bowel syndrome. Gastroenterology 2004;126:693–702.
[100] Uribe A, Alam M, Johansson O, Midtvedt T, Theodorsson E. Microflora modulates endocrine cells in gastrointestinal mucosa of the rat. Gastroenterology 1994;107: 1259–69.
[101] O'Sullivan MA, O'Morain CA. Bacterial supplementation in the irritable bowel syndrome: a randomised double-blind crossover study. Dig Liver Dis 2000;32:294–301.
[102] O'Mahony L, McCarthy J, Kelly P, et al. *Lactobacillus* and *Bifidobacterium* in irritable bowel syndrome (IBS): symptom responses and relationship to cytokine profiles. Gastroenterology 2005;128:541–51.
[103] Saggioro A. Probiotics in the treatment of irritable bowel syndrome. J Clin Gastroenterol 2004;38(suppl 2):S104–6.
[104] Kim HJ, Camilleri M, McKinzie S, et al. A randomized controlled trial of a probiotic, VSL#3, on gut transit and symptoms in diarrhoea-predominant irritable bowel syndrome. Aliment Pharmacol Ther 2003;17:895–904.
[105] Niedzielin K, Kordecki H, Birkenfeld B. A controlled, double-blind, randomised study on the efficacy of *Lactobacillus plantarum* 299 in patients with irritable bowel syndrome. Eur J Gastroenterol Hepatol 2001;13:1143–7.
[106] Nobaek S, Johansson M-L, Molin G, Ahrne S, Jeppsson B. Alteration of intestinal microflora is associated with reduction in abdominal bloating and pain in patients with irritable syndrome. Am J Gastroenterol 2000;95:1231–8.
[107] Sen S, Mullan MM, Parker TJ, Woolner JT, Tarry SA, Hunter JO. Effect of *Lactobacillus plantarum* 299v on colonic fermentation and symptoms of irritable bowel syndrome. Dig Dis Sci 2002;47:2615–20.
[108] Marteau P, Minekus M, Havenaar R, Huis in't Veld JH. Survival of lactic acid bacteria in a dynamic model of the stomach and small intestine: validation and the effects of bile. J Dairy Sci 1997;80:1031–7.
[109] Lorca GL, Wadström T, Fond de Valdez G, Ljungh Å. *Lactobacillus acidophilus* autolysins inhibit *Helicobacter pylori* in vitro. Curr Microbiol 2001;42:39–44.
[110] Canducci F, Armuzzi A, Cremonini F, et al. A lyophilised and inactivated culture of *Lactobacillus acidophilus* increases *Helicobacter pylori* eradication rates. Aliment Pharmacol Ther 2000;14:1625–9.
[111] Armuzzi A, Cremonini F, Ojetti V, et al. Effect of *Lactobacillus* GG supplementation on antibiotic-associated gastrointestinal side effects during *Helicobacter pylori* eradication therapy: a pilot study. Digestion 2001;63:1–7.
[112] Armuzzi A, Cremonini F, Bartolozzi F, et al. The effect of oral administration of *Lactobacillus* GG on associated gastrointestinal side effects during *Helicobacter* eradication. Aliment Pharmacol Ther 2001;15:163–9.
[113] Michetti P, Dorta G, Wiesel PH, et al. Effect of whey based culture supernatant of *Lactobacillus acidophilus* (jonsonii) La1 on *Helicobacter pylori* infections in humans. Digestion 1999;60:203–9.

[114] Felley CP, Corthesy-Theulaz I, Rivero JL, et al. Favourable effects of an acidified milk (*La-1*) on *Helicobacter pylori* gastritis in man. Eur J Gastroenterol Hepatol 2001;13:25–9.
[115] Cremonini F, Di Caro S, Covino M, et al. Effect of different probiotic preparations on anti-*Helicobacter pylori* therapy-related side effects: a parallel group, triple blind, placebo-controlled study. Am J Gastroenterol 2002;97:2744–9.
[116] Wang KY, Li SN, Liu CS, et al. Effects of ingesting *Lactobacillus*- and *Bifidobacterium*-containing yogurt in subjects with colonized *Helicobacter pylori*. Am J Clin Nutr 2004;80:737–41.
[117] Riordan SM, Skinner N, Nagree A, et al. Peripheral blood mononuclear cell expression of toll-like receptors and relation to cytokine levels in cirrhosis. Hepatology 2003;37: 1154–64.
[118] Kurtovic J, Ruettimann U, Adamson H, Bengmark S, Williams R, Riordan SM. Improvement in indocyanine green clearance following synbiotic treatment in cirrhosis. Gut 2003;52(Suppl):A3.
[119] Rayes N, Hansen S, Seehofer D, et al. Early enteral supply of *Lactobacillus* and fibre vs selective bowel decontamination (SBD)—a controlled trial in liver transplant recipients. Transplantation 2002;74:123–7.
[120] Rayes N, Seehofer D, Theruvath T, et al. Combined perioperative enteral supply of bioactive pre- and probiotics abolishes postoperative bacterial infections in human liver transplantation—a randomised, double blind clinical trial. Am J Transplant 2005;5: 125–30.
[121] Rayes N, Hansen S, Boucsein K, et al. Early enteral supply of fibre and lactobacilli vs parenteral nutrition—a controlled trial in major abdominal surgery patients. Nutrition 2002;18:609–15.
[122] Shanahan F. Probiotics and inflammatory bowel disease: from fads and fantasy to facts and future. Br J Nutr 2002;88(suppl):S5–9.
[123] Sartor RB. Therapeutic manipulations of the enteric microflora in inflammatory bowel diseases: antibiotics, probiotics and prebiotics. Gastroenterology 2004;126:1620–33.
[124] Sutherland LR, Hanauer S. Schmolmerich J, and the Taskforce on Clinical Trials of the International Organisation for the Study of Inflammatory Bowel Disease. Standards for trials of therapy in inflammatory bowel disease. Inflamm Bowel Dis 1997;3:277–83.
[125] Veldhuyzen van Zanten SJO, Talley NJ, Bytzer P, Klein KB, Whorwell PJ, Zinsmeister AR. Design of treatment trials for functional gastrointestinal disorders. Gut 1999;45(suppl 2):1169–77.
[126] Metchnikoff E. The prolongation of life: optimistic studies. London: William Heinemann; 1907.
[127] Eiseman B, Silen W, Bascom GS, Kauvar AJ. Fecal enema as an adjunct in the treatment of pseudomembranous enterocolitis. Surgery 1958;44:854–9.

Gastroenterol Clin N Am 34 (2005) 437–450
GASTROENTEROLOGY CLINICS
OF NORTH AMERICA

Probiotics, Gut Inflammation and Barrier Function

Erika Isolauri, MD, PhD[a,*], Seppo Salminen, PhD[b]

[a]Department of Pediatrics, University of Turku, 20520 Turku, Finland
[b]Department of Biochemistry and Food Chemistry and Functional Foods Forum, University of Turku, Turku, Finland

Mucosa of the gastrointestinal (GI) tract function as a barrier excluding and eliminating numerous antigens derived from the external environment. Protection against potentially harmful agents is ensured by numerous nonimmunologic factors, including saliva, gastric acid, peristalsis, mucus, intestinal proteolysis, and epithelial cell membranes with the intercellular junctional complexes [1]. An immunological barrier is maintained by the mucosal immune system [2]. The GI tract is home to the largest mass of lymphoid tissue in the human body, being simultaneously the site where these cells are challenged constantly. Consequently, it comprises an important element in the total immunologic capacity of the host regulating the immune responses to microorganisms and food.

The GI tract of the human adult harbors more than 500 identified species of culturable and many unculturable bacteria, some regarded as common commensals of the normal human microbiota, the complex collection and balance of microorganisms that normally inhabit the GI tract [3–5]. Their peaceful coexistence with the host is referred to as host–microbe cross-talk, implying a benefit of the microbial presence to the host. Indeed, not only is the gut microbiota an essential constituent in the intestine's defense barrier, it would appear that gut microbiota strains exercise a function in controlling and maintaining intestinal epithelial homeostasis [6,7].

The barrier function and immune regulation are linked closely and have common characteristics; most antigens are excluded by a well-functioning mucosal barrier in the gut [1]. Nevertheless, there are specialized antigen transport mechanisms in the villous epithelium and particularly in Peyer's patches, which are crucial in determining the subsequent immune responses to a given antigen [8]. Degradation of antigens is a necessary initial step in

The original research has been financially supported by the Academy of Finland (Microbes and Man Research Programme, MICMAN).

*Corresponding author. *E-mail address*: erika.isolauri@utu.fi (E. Isolauri).

0889-8553/05/$ – see front matter
doi:10.1016/j.gtc.2005.05.010

gastro.theclinics.com

controlling inflammatory responsiveness to dietary antigens [9], and most antigens encountered already are processed when they contact the mucosal surface. Proteases of specific microbiota strains contribute to the processing of antigens in the gut and modify their immunogenicity in vitro and in vivo [10–13]. The processing of dietary antigens has been linked to the potential to generate peptides with suppressive effects on lymphocyte proliferation in healthy subjects [12].

The importance of host–microbe interaction is most vital in the neonatal period, when the establishment of normal microbiota provides the host with the most substantial antigen challenge, with a strong stimulatory effect for the maturation of the gut-associated lymphoid tissue. The transition from the microbe-free conditions of intrauterine life to the extrauterine world is abrupt and drastic. In parallel, the initiation of nutrition intake by the oral route gives rise to another important source of new antigens and modifies the early gut microbiota composition. In the immature intestine, a greater amount of these antigens may traverse the mucosal barrier, and the routes of transport may be altered [1,9,11]. Mucosal dysfunction and inexperienced immune regulation may contribute to secondarily increased intestinal permeability as a result of the inflammatory response in the intestinal mucosa induced by viruses, bacteria, or dietary antigens. For this reason, impaired gut barrier function may be the explanation for an infant's proneness to allergic and infectious disease [14].

GUT MICROBIOTA—THE SOURCE OF PROBIOTICS

The intestinal microbiota does not exist as a defined entity; this population comprises a dynamic and diverse mixture of microbes. Thus, the gut microbiota of an adult generally constitute a complex community of microbes in the GI tract specific to each individual. The importance of resident bacteria for the host's physiology has been documented. Major functions of the gut microbiota include metabolic activities that result in salvage of energy and absorbable nutrients, trophic effects on the intestinal epithelium, and protection of the host against invasion by harmful microbes [5,15].

A probiotic is defined as a live microbial food supplement with a proven beneficial effect on human health. Probiotics have been selected from members of the normal healthy intestinal microbiota, most of them belonging to *Lactobacillus* or *Bifidobacterium,* but new probiotic microbes from other species and genera recently have been introduced [16].

FORMATION OF GUT MICROBIOTA

The microbiota of a newborn develops rapidly after birth and is initially markedly dependent on genetic factors, the mother's microbiota, mode of delivery, and birth environment [17,18]. Subsequently, feeding practices, formula feeding versus breastfeeding, the composition of breastmilk, in particular its two-linked fucosylated oligosaccharide (2-FOS) content, together with the living environment, influence microbiota development, at the level of species composition and number of bacteria and concentrations of different microbes [17–19].

The initial colonization process directs the later microbiota succession and health of the infant. Understanding and guidance of the gut microbiota composition appropriate for age by dietary means appear to be important future treatment targets, particularly in at-risk populations [4,20,21].

The establishment of the gut microbiota usually is characterized by specific stages of development, namely early colonization by facultative anaerobes such as enterobacteria, coliforms and lactobacilli, rapidly succeeded by anaerobic genera such as *Bifidobacterium*, *Bacteroides*, *Clostridium,* and *Eubacterium* [22]. Molecular methods indicate that lactic acid-producing bacteria may account for less than 1% of the total microbiota, while bifidobacteria can range from 60% up to 90% of the total fecal microbiota in breastfed infants. In fact, the greatest difference in the microbiota of breastfed and formula-fed infants may lie in the bifidobacterial numbers and species composition. *Bifidobacterium breve*, *B infantis,* and *B longum* are species frequently found in fecal samples from breastfed infants, while *B adolescentis* becomes more common later in life [18]. The most common lactobacilli in both breastfed and formula-fed infant feces constitute *Lactobacillus acidophilus*-group microorganisms such as *L acidophilus*, *L gasseri,* and *L johnsonii* [23]. In this phase, lactic acid bacteria may have a role in providing the right conditions for the members of the genus *Bifidobacterium* to dominate.

Following the early development and succession during the first 6 months of life, microbiota development is directed toward a more diverse community [24]. Weaning will break the contact with and constant supply of oligosaccharides and microbes from the skin of the mother. After weaning, the differences observed between breastfed and formula-fed infants disappear because of the increase in the numbers of *Bacteroides, Clostridium,* and other anaerobic cocci in the former group, with general increases in numbers of *Escherichia coli* and enterococci after weaning in both groups [25]. The numbers of *Bacteroides* and anaerobic gram-positive cocci also appear to increase gradually during and following weaning, while enterobacteria decrease [26].

After the first year of life, the shift in microbiota composition continues, and the community becomes more diverse, with *Bacteroides*, *Veillonella,* and *Fusobacterium* on the increase. At the same time, the number of unculturable microbes also increases, posing a challenge in characterizing the composition and the activity of the total microbiota. The microbiota begin to resemble that of adults, and there is a decrease in facultative anaerobes, although these remain at higher numbers than in adults [22,25]. Later there may be a decrease in the levels of clostridia with a concomitant increase in a more diverse anaerobic microbiota including microorganisms such as fusobacteria and eubacteria [27]. In contrast, other studies [28] report that in some 10- to 18-month-old infants, bifidobacteria still predominate, followed by *Bacteroides*, enterobacteria, and enterococci. Moreover, it has been reported that children (16 months to 7 years) still may harbor higher levels of bifidobacteria and enterobacteria than adults [29]. In adults, members of the genus *Bacteroides* comprise approximately 25% of the microbiota. A prominent member is *Bacteroides thetaiotaomicron*, which has specific sensors involved in the acquisition and processing of dietary polysaccharides. This adaptation may contribute to use and energy harvest from the diet [30]. The most prominent change from infancy to adulthood is the increase in the numbers of unculturable strains, so that the diversity of the developing microbiota is enhanced.

PROMOTING GUT BARRIER FUNCTION BY PROBIOTICS: POTENTIAL FOR CLINICAL APPLICATION

Gut microbiota as treatment target: probiotics versus prebiotics

Specific aberrancies in the intestinal microbiota may predispose the host to disease. The composition and concentration of *Bifidobacterium* and *Clostridium* have been used as biomarkers for the gut microbiota, and alterations have been suggested in allergic, inflammatory, and infectious diseases [9,21,31]. Intestinal mucosal defense mechanisms acting in lumen and mucosa restrict colonization by pathogenic bacteria by interfering with the adherence of microorganisms to the mucosal surface. Such colonization resistance also may take place through competition for nutrients and attachment space and modified metabolic activity of the microbiota hindering the survival of entering foreign bacteria. In this way, normal gut microbiota can prevent the overgrowth of potential pathogens in the GI tract. One of the selection criteria for probiotics is the production of antimicrobial substances [32].

Probiotics are live microbial food supplements or components of bacteria which have been demonstrated to have beneficial effects on human health [33]. Probiotics are members of the healthy gut microbiota and assist in mimicking the healthy microbiota of healthy subjects. The concept of probiotics is presented in Fig. 1. Prebiotics are defined as nonabsorbable food components that beneficially stimulate one or more of the gut-beneficial microbe groups and thus have a positive effect on human health [33]. The prerequisite for this activity is that such strains are already available for stimulation in the gut. The prebiotic concept is illustrated in Fig. 1.

Probiotics introduce new microbes to the GI tract to enhance microbiota maintenance and modification, while most prebiotic components have been shown to enhance the *Bifidobacterium* biota. Each prebiotic component possesses different microbiota-modifying properties [34]. Most prebiotics identified are nondigestible oligosaccharides, and those containing fructose (inulin, oligofructose, fructo-oligosaccharides) are used most commonly. Galacto-oligosaccharides increasingly have been investigated, however, especially for use in infant formulas. It recently was reported that some fructo-oligosaccharides enhance the levels of unknown microbes in the human gut. This may be one factor potentially facilitating changes related to untoward effects [35].

Probiotics hitherto have been selected using very few, mainly technological criteria, with the recognition that some indigenous bacteria can be pathogenic, while some species are not linked to a disease risk [36]. Currently, selection

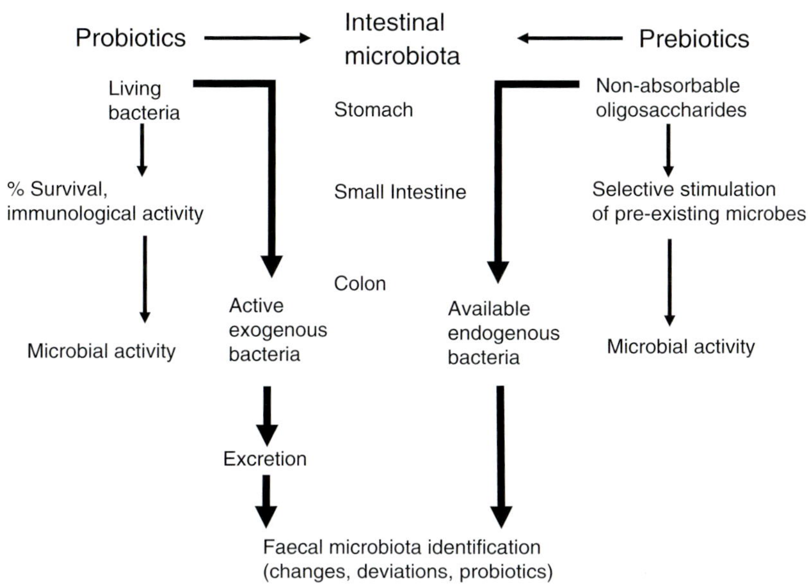

Fig. 1. The probiotic and prebiotic concepts: influencing intestinal microbiota by viable bacterial supplements versus nonabsorbable bacterial substrates.

is based on the healthy human intestinal or mucosal microbiota as the main source of new strains, with added emphasis on survival in the gut, acid and bile stability, and temporal colonization of the mucosal surfaces in the intestinal tract [21,33]. The most frequently used genera qualifying as probiotics on this selection criterion are lactobacilli and bifidobacteria.

Although many of the health-promoting aspects of probiotics have yet to be definitively proven in people, the current applications are based on clinical effects of clearly defined strains. There are numerous human intervention studies assessing the efficacy of specific probiotics in treatment and risk reduction in infectious and antibiotic-associated diarrhea [4,21]. The most fully documented probiotic intervention is the treatment of acute rotavirus diarrhea in children. Few clinical effects have been reported for prebiotics apart from intestinal microbiota modification. Indeed, eliciting a change away from gut microbiota dominated by potentially harmful bacteria toward more benign microbiota, and thereby altering the intestinal milieu, may not provide the host with health-promoting attributes.

The immunomodulatory potential of probiotics [37] has opened up new angles on the role of the gut microbiota in health and disease (see Fig. 1). These advances have introduced new potential therapeutic strategies for combating allergic, infectious, and inflammatory conditions. Normalization of the properties of an unbalanced indigenous microbiota consequently may be seen more broadly as restoration of the gut immunological barrier.

Gut barrier as treatment target

Probiotics have been shown to amplify the gut mucosal barrier functions (Table 1). *L rhamnosus* GG (ATCC 53,103) and *L plantarum* 299v adhere effectively to HT 29 intestinal epithelial cells and are able to increase expression of intestinal mucins [38]. The close association between gut barrier integrity and immune response first was demonstrated in a suckling rat model [11], and *L* GG was shown to reverse the increased intestinal permeability induced by cow's milk. Before weaning, the rats were gavaged daily with cow's milk (milk group), *L* GG with milk (milk-GG group), or water (control). The transport rates of horseradish peroxidase (HRP, a macromolecular tracer with a structure corresponding to that of dietary antigens) across patch-free jejunal segments and segments containing Peyer's patches were studied at the time when gut closure normally occurs. High intact HRP absorption was detected in the milk group as compared with milk-GG group and controls. There was a trend to increased proportional transport of HRP across Peyer's patches in the milk-GG group, and the numbers of specific antibody-secreting cells against β-lactoglobulin were increased significantly in the milk-GG group as compared with the milk group and controls.

Nonspecific use of probiotics has the aim of dampening inflammation in the gut. The mechanisms here may involve anti-inflammatory mediators generated in the interaction of the gut mucosal barrier with the intraluminal bacteria. The innate immune system is able to distinguish microbial antigens through molecules such as toll-like receptors (TLR) and CD14, which recognize conserved pathogen-associated molecular patterns not unique to pathogens but shared by commensals [39]. Thus far, it is not understood which mechanisms reduce the inflammatory response to commensals constantly present in the healthy intestine. In fact, data are accumulating to suggest that the establishment of indigenous microbiota impacts on healthy immunophysiological regulation in the gut [14]. Furthermore, it may be suggested that constant stimulation of the intestinal immune system by the indigenous

Table 1
Probiotics: impact on gut barrier function

Effect	Outcome
Exclusion and elimination of intraluminal antigens	Colonization resistance Induction of antigen degradation Reduction in the antigen load
Controlled transfer of dietary antigens	Partial degradation of dietary antigens Antigen-specific immunity
Regulation of immune response	Induction of IgA antibodies Generation of anti-inflammatory cytokines
Host–microbe cross-talk	Establishment of colonization resistance Maturation of the gut barrier functions Local immunologic homeostasis

intestinal microbiota promote the anti-inflammatory tone of the gut barrier function.

One explanation is the IgA-mediated mucosal immune response triggered by specific gut microbiota bacteria. Probiotics have been shown to stimulate nonspecific host resistance to microbial pathogens (eg, macrophage activity) and thereby aid in their eradication and immune elimination [40]. Oral introduction of specific microbiota bacteria can increase systemic and mucosal IgA responses [41–43]. The clinical correlate of this effect is promotion of the intake of oral vaccines and reduction of the risk of infection [4]. Intestinal dendritic cells can sample microbes and other intestinal antigens, even though most are killed by macrophages. Maturation of dendritic cells carrying commensals and subsequent secretion of cytokines and chemokines then influence the polarization of T helper cells and thereby the adaptive immune responses, ensuring a local IgA response [44]. This type of immune response has been suggested to prevent commensals from breaching the gut mucosal barrier, while pathogenic bacteria preferably destroy it [11,44,45]. A recent series of experimental studies in mice deficient in MyD88, an adaptor molecule essential for the TLR-mediated induction of inflammatory cytokines, or TLR2 and TLR4, demonstrated that this signaling controls the homeostasis of the epithelium and appears critical for protection of the host against gut injury in controlling cytoprotective factors and epithelial cell proliferation [46].

Another explanation may be a direct prokaryotic anti-inflammatory effect. Microbial activation of the immune system through TLRs is mediated by inducing the transcription factor NF-κB pathway [39]. Nonpathogenic enteric microbes, however, have been observed to elicit an immunosuppressive effect on intestinal epithelial cells by inhibiting the same pathway [47].

Aberrant gut microbiota may underlie not only nonspecific GI symptoms or acute infections, but also chronic diseases in inducing or perpetuating immuno-inflammatory responses. Specific use of probiotics involves modulation of the host's immune responses to potentially harmful antigens. Bovine casein degraded with *L rhamnosus* GG-derived proteases has been shown to suppress lymphocyte proliferation dose-dependently and reduce anti-CD3-stimulated interleukin (IL)-4 production by peripheral blood mononuclear cells in children with atopic eczema [12,13]. Homogenates derived from *B lactis*, *L acidophilus*, *L delbrueckii* subspecies *bulgarius* also were able to suppress proliferation of human mononuclear cells in vitro, although less effectively than *L* GG [48]. Bovine casein degraded with *L* GG-derived proteases has been found to suppress T cell activation by inhibiting protein kinase C activation [49]. Activation of protein kinase C is a necessary initial step in the NF-κB activation cascade [50].

These studies taken together, the immunologic and nonimmunologic defense mechanisms on one hand and the nonspecific and specific immunoregulatory effects on the other, act cooperatively, explaining the redundancy of immune regulation. The paradoxical regulation of the immune response to enteral antigens, triggering of the innate and adaptive immune response and oral tolerance, appears to be a further attribute of the gut immune defense.

Oral tolerance has been defined as the induction of peripheral unresponsiveness as a result of oral administration of soluble protein antigen [51], or, more recently, as any mechanism by which a potentially injurious immune response is prevented, suppressed, or shifted to a noninjurious class of immune response [52].

Tolerance may be achieved through several distinct mechanisms, including the generation of a unique subset of T lymphocytes designated T helper (Th) 3 and T regulatory cells, producing suppressive cytokines such as IL-10 and transforming growth factor (TGF)-β, respectively. These cells actively control or suppress the function of other cells, including Th1 and Th2 cells, and thus have been implicated in protection against Th1-type autoimmune disease and Th2-mediated atopic disease.

The role of the intestinal microbiota in oral tolerance induction to the IgE response first was demonstrated in germ-free mice [53]. In contrast to control mice, germ-free animals maintained their tendency to systemic immune responses such as production of IgE antibodies upon oral administration of ovalbumin. Abrogation of oral tolerance was caused by lack of intestinal microbiota. The aberrant IgE response in germ-free mice could be corrected by microbiota reconstitution at the neonatal stage, but not later. Mice lacking a functional receptor for TLR4 showed heightened susceptibility to allergic responses to food antigens [54]. A strain of *L casei* has been demonstrated to suppress IgE responses and systemic anaphylaxis in a murine model of food allergy [55], and in allergic contact dermatitis and T cell-mediated antigen-specific skin inflammation [56].

To evaluate the clinical effect of probiotic therapy in food allergy, infants with atopic dermatitis and challenge-proven cow's milk allergy were fed an extensively hydrolyzed whey formula or a similar formula containing *L* GG [57]. There was a significant improvement in the clinical course of atopic dermatitis concomitant with a reduction in the concentrations of tumor necrosis factor-α during the 1-month management with probiotics. The potential of probiotics to control allergic inflammation at an early age has been assessed in a randomized double-blind placebo-controlled study [58]. Infants manifesting atopic eczema during exclusive breastfeeding were given probiotic-supplemented, *B lactis* Bb-12 or *L* GG, extensively hydrolyzed whey formulas, or the same formula without probiotics when formula feedings were necessitated. In parallel with an improvement in skin condition in patients receiving probiotic-supplemented formulas as compared with the unsupplemented group, there was a reduction in the concentration of soluble CD4 in serum and eosinophilic protein X in urine after 2 months of therapy. The total duration of the intervention was 6 months. Similar clinical results with probiotic intervention recently were obtained in older children with the condition [59].

In health, tolerance also is generated and maintained to normal microbiota, and such tolerance may be abrogated in patients with inflammatory bowel disease [60]. Preliminary reports indicate benefit from probiotic intervention in reversing some of the immunologic disturbances, and in lessening disease activity and increasing intestinal permeability in pediatric patients with Crohn's disease [4,21,61]. In adults undergoing operation for the condition, however, *L* GG failed to prevent endoscopic recurrence during 1 year of follow-up [62]. A recent study provides evidence supporting treatment with a nonpathogenic *E coli* in maintaining remission in ulcerative colitis [63]. Results of a clinical trial in adults using a preparation containing four strains of lactobacilli (*L casei, L plantarum, L acidophilus, and L delbrückii* subspecies *bulgaricus*) and three bifidobacteria strains (*B longum, B breve,* and *B infantis*) together with *Streptococcus salivarius* subspecies *thermophilus* were encouraging in respect of preventing relapses of chronic pouchitis [64,65].

These early clinical reports show constant benefit of probiotics in children with infectious and allergic conditions, but not in adults. Treatment trials in adults have applied infant-type probiotics, so that the discrepancy may be because of different composition of the normal gut microbiota at different ages, as interaction between probiotic and resident bacteria occurs [5,14,20,33]. Furthermore, some probiotic strains adhere better to the small intestine, while some bind specifically to different parts of the large intestine [66]. It is likely that strains also adhere differently in healthy versus damaged mucosa [67], and even closely related probiotics have been shown to possess different in vitro properties [68], this possibly also explaining differences in clinical effects. The distinction may be explained in part by the recently demonstrated difference between children and adults in responding to bacterial products [69], underlining the importance of the maturation of the immune system also.

The immature gut barrier as treatment target

The host–microbe interaction, in terms of the generation of immunophysiological regulation, is particularly apparent during postnatal development, when major maturational events occur (eg, appearance of mucosal proteins, digestive enzymes, and the development of the intestinal microbiota). It has been demonstrated in experimental animal models that the capacity to generate IgA-producing cells is initiated with the establishment of the gut microbiota. Additionally, with the onset of a specific IgA response, the number of translocating bacteria drops, reflecting maturation of the intestine's immunologic defense mechanisms [41].

Vaginally born infants and infants born by caesarean section showed major differences in culturable microbiota, still to be observed at 6 months of age [42]. Colonization appeared to be associated with the maturation of humoral immune mechanisms. *Bacteroides fragilis* and, to a lesser extent, *Bifidobacterium* species were important in this respect, as infants harboring these organisms had more circulating IgA- and IgM-secreting cells.

These results suggest that the intestinal microbiota is important in immune regulation in human individuals, and that qualitative differences in the composition of the gut microbiota might affect the immunologic homeostasis and subsequent risk of disease. The intestinal microbiota from infants at high risk of atopic diseases was analyzed at 3 weeks of age by conventional bacterial cultivation and two culture-independent methods [31]. A positive skin prick reaction at 12 months was observed in 29% of the subjects. At 3 weeks of age, the bacterial cellular fatty acid profile in fecal samples differed between those infants who later manifested atopic sensitization and those who did not. Fluorescence in situ hybridization was used to show that atopic subjects have more *Clostridium* species and fewer *Bifidobacterium* species in stools compared with nonatopics. The preventive potential of probiotics in atopic disease has been demonstrated in a double-blind, placebo-controlled study [70,71]. Probiotics administered pre- and postnatally for 6 months to children at high risk of atopic disease reduced the prevalence of atopic eczema to half (23%) as compared with infants receiving placebo (46%). When probiotic supplementation was given to the lactating mother, the amount of TGF-β in breast milk could be promoted. In view of the capacity of this cytokine to promote specific IgA production [72], the study would indicate one mechanism by which the infant's gut barrier function could be strengthened and links the different immunological effects of probiotics. These include stimulation of a nonspecific immune response, control of the balance between pro-and anti-inflammatory cytokines, and strengthening of the gut barrier (see Table 1).

FUTURE PERSPECTIVES FOR PROBIOTICS

The range of options available to characterize the intestinal microbiota has expanded dramatically with the advent of molecular techniques. Most of the current knowledge of the intestinal microbiota has been derived from studies using the culture method and fecal samples. Real-time quantitative polymerase chain reaction procedures are among the promising tools for studies on intestinal microbiota composition. The development of DNA microarray methodology for intestinal cells and intestinal microbiota members is fundamentally changing the understanding of microbiota–host interactions. This development needs to be enhanced by assessment of microbiota in different parts of the intestine and mucosal surfaces. This will require new sampling techniques.

The capacity to assess the gut microbiota has progressed in parallel with the advances in host–microbe cross-talk. Such understanding will lead to the development of a new generation of probiotics, the action of which could be selected for defined disease-associated deviations detected in new microbiota studies. Such studies also will facilitate the potential use of genetically modified probiotic bacteria for pharmaceutical uses, as indicated by the initial studies on *L lactis* engineered to produce IL-10 locally [73].

Numerous probiotics have a long history of safe use, and no health concerns have been observed [33]. New probiotic strains lacking such a history need to be assessed for their safety on a strain-by-strain basis. Safety assessment procedures have been developed for novel microbes, including probiotics and genetically modified probiotics.

SUMMARY

Probiotic supplementation aims to provide safe yet sufficient microbial stimulus for the host immune system and intestinal microbiota by means of cultures of beneficial live microorganisms characteristic of the healthy infant gut microbiota, and to contribute to the anti-inflammatory tone of the intestinal milieu. The aims of intervention are to avert deviant microbiota development, impaired gut barrier function, abnormal immune responsiveness, and immuno-inflammatory disease. Pre-clinical testing of candidate probiotics requires new criteria applied in selection, assessment of safety, and evaluation of markers of physiological function. Scientific efforts for validating future probiotics should be directed to elucidate the characteristics of distinct probiotic strains, their effects on gut barrier function, and the immunologic mechanisms of action. Rigorous clinical documentation of probiotic strains will be required for developing clinical recommendations to reduce disease risk in target populations.

References

[1] Sanderson IR, Walker WA. Uptake and transport of macromolecules by the intestine: possible role in clinical disorders (an update). Gastroenterology 1993;104:622–39.
[2] Brandtzaeg P. Molecular and cellular aspects of the secretory immunoglobulin system. APMIS 1995;103:1–19.
[3] Berg RD. The indigenous gastrointestinal microflora. Trends Microbiol 1996;4:430–5.
[4] Cummings JH, Antoine JM, Aspiroz F, et al. PASSCLAIM (process for the Assessment of Scientific Support for Claims on Foods). Gut health and immunity. Eur J Nutr 2004;43 (Suppl 2):118–73.
[5] Guarner F, Malagelada JR. Gut flora in health and disease. Lancet 2003;361:512–9.
[6] Gaskins HR. Immunological aspects of host/microbiota interactions at the intestinal epithelium. In: Mackie RI, White BA, Isaacson RE, editors. Gastrointestinal microbiology. New York: International Thomson Publishing; 1997. p. 537–87.
[7] Cebra JJ. Influences of microbiota on intestinal immune system development. Am J Clin Nutr 1999;69:1046–51.
[8] Ducroc R, Heyman M, Beaufrere B, et al. Horseradish peroxidase transport across rabbit jejunum and Peyer's patches in vitro. Am J Physiol 1983;245:G54–8.
[9] Isolauri E, Rautava S, Kalliomäki M, et al. Role of probiotics in food hypersensitivity. Curr Opin Allergy Clin Immunol 2002;2:263–71.
[10] Barone KS, Reilly MR, Flanagan MP, et al. Abrogation of oral tolerance by feeding encapsulated antigen. Cell Immunol 2000;199:65–72.
[11] Isolauri E, Majamaa H, Arvola T, et al. *Lactobacillus casei* strain GG reverses increased intestinal permeability induced by cow milk in suckling rats. Gastroenterology 1993;105:1643–50.
[12] Sütas Y, Soppi E, Korhonen H, et al. Suppression of lymphocyte proliferation in vitro by bovine caseins hydrolysed with *Lactobacillus casei* GG-derived enzymes. J Allergy Clin Immunol 1996;98:216–24.
[13] Sütas Y, Hurme M, Isolauri E. Down-regulation of antiCD3 antibody-induced IL-4 production by bovine caseins hydrolysed with *Lactobacillus* GG-derived enzymes. Scand J Immunol 1996;43:687–9.
[14] Rautava S, Ruuskanen O, Ouwehand A, et al. The hygiene hypothesis of atopic disease- an extended version. J Pediatr Gastroenterol Nutr 2004;38:378–88.
[15] Backhed F, Ding H, Wang T, et al. The gut microbiota as an environmental factor that regulates fat storage. Proc Natl Sci U S A 2004;101:15718–23.

[16] Salminen S, Ouwehand A, Benno Y, et al. Probiotics: how should they be defined? Trends in Food Science & Technology 1999;10:107–10.
[17] Favier C, Vaughan E, de Vos W, et al. Molecular monitoring of succession of bacterial communities in human neonates. Appl Environ Microbiol 2002;68:219–26.
[18] Benno Y, Mitsuoka T. Development of intestinal microflora in humans and animals. Bifidobacteria Microflora 1986;5:13–25.
[19] Newburg DS, Ruiz-Palacios GM, Altaye M, et al. Human milk alpha 1, 2-linked fucosylated oligosaccharides decrease risk of diarrhea due to stable toxin of *E coli* in breastfed infants. Adv Exp Med Biol 2004;554:457–61.
[20] Edwards CA, Parrett AM. Intestinal flora during the first months of life: new perspectives. Br J Nutr 2002;88(Suppl 1):S11–8.
[21] Isolauri E, Kirjavainen PV, Salminen S. Probiotics—a role in the treatment of intestinal infection and inflammation? Gut 2002;50(Suppl 3):S54–9.
[22] Harmsen HJ, Wildeboer-Veloo AC, Raangs GC, et al. Analysis of intestinal flora development in breast-fed and formula-fed infants by using molecular identification and detection methods. J Pediatr Gastroenterol Nutr 2000;30:61–7.
[23] Vaughan E, de Vries M, Zoetendal E, et al. The intestinal LABs. Antonie Van Leeuwenhoek 2002;82:341–52.
[24] Mikelsaar M, Mändar R, Sepp E, et al. Human lactic acid microflora and its role in the welfare of host. In: Salminen S, von Wright A, Ouwehand A, editors. Lactic acid bacteria: microbiology and functional aspects. New York: Marcel Dekker Incorporated; 2004. p. 456–506.
[25] Wang M, Ahrne S, Antonsson M, et al. T-RFLP combined with principal component analysis and 16S rRNA gene sequencing: an effective strategy for comparison of fecal microbiota in infants of different ages. J Microbiol Methods 2004;59:53–69.
[26] Stark PL, Lee A. The microbial ecology of the large bowel of breastfed and formula-fed infants during the first year of life. J Med Microbiol 1982;15:189–203.
[27] Langhendries JP, Paquay T, Hannon M, et al. Acquisition de la flore intestinale néonatale: rôle sur la morbidité et perspectives thérapeutiques. Arch Pediatr 1998;5:644–53.
[28] Guerin-Danan C, Chabanet C, Pedone C, et al. Milk fermented with yogurt cultures and *Lactobacillus casei* compared with yogurt and gelled milk: influence on intestinal microflora in healthy infants. Am J Clin Nutr 1998;67:111–7.
[29] Hopkins MJ, Sharp R, Macfarlane GT. Age and disease related changes in intestinal bacterial populations assessed by cell culture, 16S rRNA abundance, and community cellular fatty acid profiles. Gut 2001;48:198–205.
[30] Xu J, Bjursell MK, Himrod J, et al. A genomic view of the human–*Bacteroides thetaiotaomicron* symbiosis. Science 2003;299:2074–6.
[31] Kalliomäki M, Kirjavainen P, Eerola E, et al. Distinct patterns of neonatal gut microflora in infants in whom atopy was and was not developing. J Allergy Clin Immunol 2001;107:129–34.
[32] Ouwehand AC, Salminen S, Isolauri E. Probiotics: an overview of beneficial effects. Antonie Van Leeuwenhoek 2002;82:279–89.
[33] Salminen S, Bouley C, Boutron-Ruault MC, et al. Functional food science and gastrointestinal physiology and function. Br J Nutr 1998;80(Suppl 1):147–71.
[34] Gibson GR. Dietary modulation of the human gut microflora using prebiotics. Br J Nutr 1998;80:S209–12.
[35] Apajalahti JH, Kettunen H, Kettunen A, et al. Culture-independent microbial community analysis reveals that inulin in the diet primarily affects previously unknown bacteria in the mouse cecum. Appl Environ Microbiol 2002;68:4986–95.
[36] Sullivan A, Nord CE. Probiotics and gastrointestinal diseases. J Intern Med 2005;257:78–92.
[37] Isolauri E, Sütas Y, Kankaanpää P, et al. Probiotics: effects on immunity. Am J Clin Nutr 2001;73:S444–50.

[38] Mack DR, Michail S, Wel S, et al. Probiotics inhibit enteropathogenic *E coli* adherence in vitro by inducing intestinal mucin gene expression. Am J Physiol 1999;276:G941–50.
[39] Zhang G, Ghosh S. Toll-like receptor-mediated NF-kappaB activation: a phylogenetically conserved paradigm in innate immunity. J Clin Invest 2001;107:13–9.
[40] Perdigón G, Nader de Macías ME, Alvarez S, et al. Systemic augmentation of the immune response in mice by feeding fermented milks with *Lactobacillus casei* and *Lactobacillus acidophilus*. Immunology 1988;63:17–23.
[41] Shroff KE, Meslin K, Cebra JJ. Commensal enteric bacteria engender a self-limiting humoral mucosal immune response while permanently colonizing the gut. Infect Immun 1995; 63:3904–13.
[42] Grönlund MM, Arvilommi H, Kero P, et al. Importance of intestinal colonisation in the maturation of humoral immunity in early infancy: a prospective follow up study of healthy infants aged 0–6 months. Arch Dis Child 2000;83:F186–92.
[43] Kaila M, Isolauri E, Soppi E, et al. Enhancement of the circulating antibody secreting cell response in human diarrhea by a human lactobacillus strain. Pediatr Res 1992; 32:141–4.
[44] Macpherson AJ, Uhr T. Induction of protective IgA by intestinal dendritic cells carrying commensal bacteria. Science 2004;303:1662–5.
[45] Isolauri E, Gotteland M, Heyman M, et al. Antigen absorption in rabbit bacterial diarrhea (RDEC-1). In vitro modifications in the ileum and Peyer's patches. Dig Dis Sci 1990;35: 360–7.
[46] Rakoff-Nahoum S, Paglino J, Eslami-Varzaneh F, et al. Recognition of commensal microflora by toll-like receptors is required for intestinal homeostasis. Cell 2004;118: 229–41.
[47] Neish AS, Gewirtz AT, Zeng H, et al. Prokaryotic regulation of epithelial responses by inhibition of IkappaB-alpha ubiquitination. Science 2000;289:1560–3.
[48] Pessi T, Sütas Y, Saxelin M, et al. Antiproliferative effects of homogenates derived from five strains of candidate probiotic bacteria. Appl Environ Microbiol 1999;65:4725–8.
[49] Pessi T, Isolauri E, Sütas Y, et al. Suppression of T cell activation by *Lactobacillus* GG-degraded bovine casein. Int Immunopharmacol 2001;1:211–8.
[50] Tan SL, Parker PJ. Emerging and diverse roles of protein kinase C in immune cell signaling. Biochem J 2003;376(Pt 3):545–52.
[51] Nagler-Anderson C. Tolerance and immunity in the intestinal system. Crit Rev Immunol 2000;20:103–20.
[52] Weiner HL. Oral tolerance: immune mechanisms and the generation of Th3-type TGF-beta-secreting regulatory cells. Microbes Infect 2001;3:947–54.
[53] Sudo N, Sawamura S, Tanaka K, et al. The requirement of intestinal bacterial flora for the development of an IgE production system fully susceptible to oral tolerance induction. J Immunol 1997;159:1739–45.
[54] Bashir MEH, Louie S, Shi HA, et al. Toll-like receptor 4 signalling by intestinal microbes influences susceptibility to food allergy. J Immunol 2004;172:6978–87.
[55] Shida K, Takahashi R, Iwadate E, et al. *Lactobacillus casei* strain Shirota suppresses serum immunoglobulin E and immunoglobulin G1 responses and systemic anaphylaxis in a food allergy model. Clin Exp Allergy 2002;32:563–70.
[56] Chapat L, Chemin K, Dubois B, et al. *Lactobacillus casei* reduces $CD8^+$ T cell-mediated skin inflammation. Eur J Immunol 2004;34:2520–8.
[57] Majamaa H, Isolauri E. Probiotics: a novel approach in the management of food allergy. J Allergy Clin Immunol 1997;99:179–85.
[58] Isolauri E, Arvola T, Sutas Y, et al. Probiotics in the management of atopic eczema. Clin Exp Allergy 2000;30:1604–10.
[59] Rosenfeldt V, Benfeldt E, Nielsen SD, et al. Effect of probiotic *Lactobacillus* strains in children with atopic dermatitis. J Allergy Clin Immunol 2003;111:389–95.

[60] Duchmann R, Kaiser I, Hermann E, et al. Tolerance exists towards resident intestinal flora but is broken in active inflammatory bowel disease (IBD). Clin Exp Immunol 1995;102: 448–55.
[61] Sartor RB. Therapeutic manipulation of the enteric microflora in inflammatory bowel diseases: antibiotics, probiotics, and prebiotics. Gastroenterology 2004;126:1620–33.
[62] Prantera C, Scribano ML, Falasco G, et al. Ineffectiveness of probiotics in preventing recurrence after curative resection for Crohn's disease: a randomised controlled trial with *Lactobacillus* GG. Gut 2002;51:405–9.
[63] Rembacken BJ, Snelling AM, Hawkey PM, et al. Nonpathogenic *Escherichia coli* versus mesalazine for the treatment of ulcerative colitis: a randomised trial. Lancet 1999; 354:635–9.
[64] Gionchetti P, Rizzello F, Venturi A, et al. Oral bacteriotherapy as maintenance treatment in patients with chronic pouchitis: a double-blind, placebo-controlled trial. Gastroenterology 2000;119:305–9.
[65] Gionchetti P, Rizzello F, Helwig U, et al. Prophylaxis of pouchitis onset with probiotic therapy: a double-blind, placebo-controlled trial. Gastroenterology 2003;124:1202–9.
[66] Ouwehand AC, Salminen S, Tolkko S, et al. Resected human colonic tissue: new model for characterizing adhesion of lactic acid bacteria. Clin Diagn Lab Immunol 2002;9:184–6.
[67] Ouwehand AC, Salminen S, Roberts PJ, et al. Disease-dependent adhesion of lactic acid bacteria to the human intestinal mucosa. Clin Diagn Lab Immunol 2003;10:643–6.
[68] Ouwehand A, Salminen S. In vitro adhesion assays for probiotics and their in vivo relevance. Microb Ecol Health Dis 2003;15:175–84.
[69] Tulic MK, Fiset PO, Manoukian JJ, et al. Role of toll-like receptor 4 in protection by bacterial lipopolysaccharide in the nasal mucosa of atopic children but not adults. Lancet 2004;363:1689–97.
[70] Kalliomäki M, Salminen S, Arvilommi H, et al. Probiotics in primary prevention of atopic disease: a randomised placebo-controlled trial. Lancet 2001;357:1076–9.
[71] Kalliomäki M, Salminen S, Poussa T, et al. Probiotics and prevention of atopic disease: 4-year follow-up of a randomised placebo-controlled trial. Lancet 2003;361:1869–71.
[72] Rautava S, Kalliomäki M, Isolauri E. Probiotics during pregnancy and breastfeeding might confer immunomodulatory protection against atopic disease in the infant. J Allergy Clin Immunol 2002;109:119–21.
[73] Steidler L, Hans W, Schotte L, et al. Treatment of murine colitis by *Lactococcus lactis* secreting interleukin-10. Science 2000;289:1352–5.

Gastroenterol Clin N Am 34 (2005) 451–463

GASTROENTEROLOGY CLINICS
OF NORTH AMERICA

Pediatric Applications of Probiotics

Jon A. Vanderhoof, MD[a,b], Rosemary J. Young, MS, RN[a,*]

[a]University of Nebraska Medical Center, 985160 Nebraska Medical Center, Omaha, NE 68198, USA

[b]Global Medical Affairs, Mead Johnson Nutritionals, 2400 West Lloyd Expressway, Evansville, IN 47721, USA

The concept of probiotics was established early in the last century at the Pasteur Institute by the work of Metchnikoff, who hypothesized that the ingestion of fermented milk products had a beneficial impact on the health and lifespan of Bulgarian peasants [1]. The formal term probiotic first was referenced in 1954 by Ferdinand Vergin in an article discussing the effects of antibiotics on beneficial intestinal bacteria [2]. Fuller often is credited with establishing the first definition of probiotics as a bacterial product inferring benefits based on his work with animals [3].

The accepted definition of a probiotic encompasses the cumulative state of the research in this area. A probiotic is a beneficial species of bacteria that colonizes and replicates in the human intestinal tract and provides a positive benefit to the host. Pediatric applications of probiotics are widespread, with the primary applications being diarrhea and allergic disorders. This article reviews normal infant colonization, how native intestinal bacteria influence health, and research regarding probiotic use in infants and children.

COLONIZATION

At birth, the intestine is sterile. The digestive tract provides an excellent environment for bacterial growth, with minimal oxygen and plentiful nutrients. This nutrient-rich aqueous environment is an ideal culture medium for bacteria. Additionally, the gut epithelium itself is lined with mucins, extensively glycosylated proteinaceous compounds that function as binding sites for many of the gut bacteria and function to assist or inhibit bacterial attachment to the underlying epithelium or antigen sampling cells. Some probiotics are capable of upregulating mucin genes in the gut and thereby altering bacterial flora [4].

The establishment of native bacteria that are able to thrive in the infant's developing intestinal tract without provoking an adverse immunological response is a complex process that is completed in the first year and remains relatively stable throughout life [5]. The establishment of native flora creates

*Corresponding author. *E-mail address*: rjyoung@unmc.edu (R.J. Young).

0889-8553/05/$ – see front matter
doi:10.1016/j.gtc.2005.05.001

gastro.theclinics.com

the basis for the gut-associated lymphatic system to provide a noninflammatory environment that acts as a barrier against foreign or pathogenic bacteria. The infant's gut undergoes a continuous process of immune sampling of the intestinal contents. This process occurs at specific binding sites or receptors on the surface of the gastrointestinal (GI) organisms that interact with antigen sampling cells in or below the gut epithelium in a very complex fashion to establish tolerance [6]. The bacteria that then are recognized as native are able to inhibit the same inflammatory pathway that pathogenic organisms elicit to cause disease. In the infant, immune sampling begins with predominately large numbers of gram-positive bacteria, particularly lactic acid bacteria, including lactobacilli, streptococci, enterococci, lactococcus, and bifidobacteria. Distinct strains and species of the various bacteria then permanently are recognized as native by that particular infant.

Colonization begins immediately at birth through complex interactions with the environment, initially through exposure to maternal flora. Factors affecting maternal vaginal and fecal flora are determined by the mother's own birth exposures and overall maternal health through the later weeks of pregnancy. Vaginally delivered infants initially have significantly different flora than infants born by cesarean section [7]. The immediate birth and home environmental exposures in the subsequent days and weeks create further refinement of the infant's intestinal microbiota.

After birth exposure, nondigestible food ingredients that stimulate the growth or activity of bacteria direct much of the infant's colonization. These nutrients consumed by the infant consist primarily of carbohydrates such as fibers. Nonabsorbable oligosaccharides such as galacto–oligosaccharides, fructo–oligosaccharides, gluco–oligosaccharides and xylo–oligosaccharides enhance growth of certain strains of bacteria and as such are known as prebiotics. Colostrum contains a high level of fructo–oligosaccharides that appear to enhance the selective bacterial colonization of the breastfed infant with bifidobacteria [8]. The recent addition of various types of prebiotics to infant formulas has resulted in the development of colonic flora more closely resembling that of infants fed human milk [9]. Some fructo–oligosaccharides have been shown in animal studies to potentially have harmful effects [10], suggesting that care should be used in the selection of the most appropriate product. Nucleotides and gangliosides also have been shown to enhance bifidobacteria growth and have been evaluated as beneficial additives to infant formulas.

At birth, facultative anaerobes such as enterobacteria, coliforms, and lactobacilli are established. Initially both breast- and bottle-fed infants have some similar flora, including *Lactobacillus acidophilus*, *L gasseri*, and *L johnsonii*. Quickly, bifidobacteria are established and overall account for the greatest percentage of fecal bacteria in all infants, but they are present in higher numbers in infants fed human milk. The breastfed infant tends to have a predominance of *Bifidobacterium breve*, *B infantis* and *B longum* [11]. Formula-fed infants have more bacteroides, clostridia, and *Enterobacteriaceae* [12]; however, very little is known about infants fed a mixture of human and artificial milk.

Human milk contains immunological factors such as secretory IgA and lysozyme that influence the developing microbiota. It has been hypothesized that the flora that result from these effects in the breastfed infant provide for the establishment of greater anti-inflammatory activity against antigens than the flora of infants fed artificial milk, who lack these immunological factors [13]. This may account, in part, for why human milk is less allergenic than most cow's milk or soy whole protein formulas. Iron content of the feeding also influences the bacterial population. It has been shown that human milk, low in iron content but containing highly bioavailable iron, is more compatible with the growth of bifidobacteria and lactobacilli. Iron-supplemented formula may promote clostridia and enterobacteria growth. More recently, the addition of lactoferrin and changes in whey to casein protein ratios added to artificial formula have been shown to influence gut flora [14,15].

It has been shown that the short-chain fatty acid profiles differ between breast- and formula-fed infants until 12 months of age [16,17]. This is because infants fed human milk receive more total nondigestible oligosaccharides than infants fed formula. In both groups, the ability to handle complex carbohydrates is limited until later infancy in part because of pancreatic immaturity. Infants fed human milk have fewer gram-negative anaerobes, which contribute additional fermentation of carbohydrates traversing through the colon [18]. Parrett and Edwards have shown that formula-fed infants develop the capacity for fermentation of complex carbohydrates much earlier than breast-fed infants because of differing environmental exposures [19].

As the infant matures, changes in native bacterial flora occur, with a more diverse population appearing during weaning from breast or formula feeding. Part of this occurs because of a decrease in oligosaccharide intake. Various formula supplements and solid foods enhance the growth of obligate anaerobes contributing to these changes [20]. Antimicrobial use during early infancy also may have a long-standing impact on the establishment of native flora [21]. Alterations in what is considered normal colonization have been proposed as a causative factor in the development of disease.

In the adult, aerobic flora predominates in the upper GI tract, with strict anaerobes being found in the large bowel. Numbers of bacteria also increase proximally to distally with 10^3 bacteria/mL in the duodenum to 10^{11}/mL in the large bowel. This complex intestinal bacteria milieu is thought to number approximately 10 trillion cells containing 500 different microbial species. Many of these species cannot be cultured and are identified only by molecular techniques. Although the species can differ greatly, and some differences are based on age, people generally have the same dominant genera [22], including:

- Bacteroides
- Bifidobacteria
- Fusobacteria
- Eubacteria
- Lactobacilli
- Peptostreptococcus

- Clostridia
- Escherichia
- *Veillonella*

The primary mechanism preventing overgrowth of bacteria at all ages is intestinal peristalsis that propels the gut contents distally, forcing the remaining nutrients and the large volumes of bacteria into the colon. In addition, the largest fraction of immunoglobulins secreted in the human body, the secretory IgA originating in the GI tract, also aids in preventing bacterial proliferation [23]. Numerous resident bacteria themselves also may produce chemicals or bacteriocins that kill or inhibit multiplication of other surrounding bacterial species [24].

Indigenous bacteria are not highly pathogenic, but all can create opportunistic infections. Some *Clostridia* and *Bacteroides* species can infect immunocompromised patients particularly. Most species of lactobacilli and bifidobacteria are considered to have health-promoting benefits and aid in colonization resistance in the GI tract. This ability of bacteria in residence to restrict other bacteria from colonizing in the intestine once was thought to be the primary function of the gut bacteria. Now it is known that native bacteria serve a major role in the first year of life to stimulate the gut immune system.

MECHANISMS OF ACTION

There is great variability in survivability between different species and even strains of bacteria [25]. Typically, bacteria that are ingested alive and survive gastric acid have the greatest probiotic effects. Instability of the gut flora is thought to be associated with development of symptoms; however, is difficult to confirm this association merely by assessment of changes in fecal bacteria. Changes in specific bacterial enzymes such as azoreductase, nitroreductase, and B-glucuronidase may provide a better index of mechanistic effects [26].

Some probiotics reside on the mucus surface of the bowel, where the luminal carbohydrate concentration is the greatest. Other bacteria adhere to the enterocyte. The adherence of bacteria, also known as competitive exclusion, once was thought to be an important mode of action of probiotics. Presently, the adherence and subsequent stimulation of the gut immune system are receiving greater attention. Competition for essential nutrients, production of antimicrobial factors, and alteration of environmental conditions such as pH also may be major factors involved in the function of probiotic bacteria. Proposed mechanisms of action include:

- Interface with adhesion
- Maintain normal macrophage function
- Enhance secretatory IgA
- Facilitate toxin elimination
- Upregulate mucin genes
- Provide favorable environment for growth of other beneficial bacteria

Bifidobacteria and lactobacilli inhibit the growth of some harmful organisms through the production of lactic, acetic, and other organic acids. In the large bowel, additional functions of probiotics include carbohydrate fermentation, with the production of short-chain fatty acids, which provide anti-inflammatory action and calories, and the production of nutrients such as vitamin B12.

PEDIATRIC CLINICAL STUDIES

Diarrheal disorders

One of the major areas of demonstrable positive benefits in of probiotics is in the treatment of mild-to-moderate acute viral diarrhea [27,28]. Isolauri was one of the first to identify, in a double-blind, placebo-controlled study, the benefits of the probiotic *L caseii* subspecies rhamnous GG (LGG) for treating acute diarrhea requiring hospitalization for oral rehydration [29]. Subsequently, similar placebo-controlled studies using LGG in Pakistan [30], Karelian Republic [31], Thailand [32] and Helsinki, Finland [33], all demonstrated varying reductions in the severity and duration of the diarrheal episode. *L reuteri* also has been shown to be effective in a placebo-controlled study of acute diarrhea in hospitalized patients [34]. Majamaa [35] also was able to demonstrate enhanced immunological responses to rotavirus when children received LGG versus placebo.

Similar studies using LGG in Brazil [36]and Peru [37], and a combination product of *L acidophilus, L bulgaricus,* and *Streptococcus thermophilus* in Canada [38] demonstrated no benefit for treating acute diarrhea. It was suggested, however, that the severe degree of diarrhea with concurrent malabsorption in these groups may have contributed to their failure to respond. Even without symptomatic benefit, alterations in bacterial byproducts such as urease production can be seen with probiotic therapy [39] and suggest that further refinements in the type or amount of the probiotic may be needed.

Prevention of community-acquired, acute diarrhea has been demonstrated in placebo-controlled trials in Peru and Finland and in a large multi-center trial in Europe using LGG [27,40,41]. Prevention of nosocomial diarrhea using a combination probiotic-supplemented formula first was demonstrated in a well-designed placebo-controlled study by Saavedra and colleagues [42] using *B bifidum* and *S thermophilus.* A similar study with *B breve* and *S thermophilus* probiotic formula confirmed these findings [43]. Gianino and colleagues, however, were not able to demonstrate a benefit of using LGG in infants with nosocomial-acquired diarrhea compared with the protective effects of breastfeeding [44].

Antibiotic-associated diarrhea (AAD) occurs because of alterations in fecal flora. It seems to occur more commonly in infants and children, and if severe, can result in the development of *Clostridium difficile,* which is more common in adults. A recent meta-analysis demonstrated that LGG in children and a nonpathogenic yeast, *Saccharomyces boulardii,* in adults have been shown in well-controlled studies to prevent AAD [45]. Not all probiotics are equally effective, as demonstrated by the lack of effectiveness of a combination of

L acidophilus and *L bulgaricus* for treating children receiving amoxicillin in a double-blind, placebo-controlled trial [46]. Studies of probiotics for treating *Helicobacter pylori* gastritis have not documented benefits in eradicating the bacterium; however, LGG, *Saccharomyces boulardii,* and a mixture of *L acidophilus* and *B lactis* have demonstrated positive protective effects against the adverse effects of the antibiotics used to treat the problem [47].

Intestinal overgrowth of the anaerobic bacteria *C difficile* can result in colitis because of production of toxins by the bacteria. Although treatment with metronidazole and vancomycin are generally effective, recurrence develops in a small subset of people, especially those with compromised immune systems. Initially, uncontrolled trials demonstrated some benefit of using LGG for preventing recurrent *C difficile* in children [48]; however, subsequent larger controlled studies have not been performed. In placebo-controlled studies *Saccharomyces boulardii* used concurrently with standard treatment has shown excellent results in well-controlled studies in adults [49,50].

Immune-mediated disorders

Mechanistic studies in animal models of inflammatory bowel disease (IBD) have demonstrated some positive effects using LGG [51,52]. In an open-label trial, Gupta and colleagues were able to demonstrate a reduction in intestinal permeability in children with Crohn's disease who were given LGG [53]. A subsequent clinical trial using LGG as an adjunct to therapy for remission maintenance in children who had Crohn's disease did not demonstrate any benefit [54]. In adults who had ulcerative colitis, one placebo-controlled study demonstrated some benefit using fermented milk with *Bifidobacterium* for relapse prevention [55]. Treatment of pouchitis, which occurs after colectomy for severe ulcerative colitis, with a combination probiotic known as VSL#3, and in a more recent study with LGG, has been shown to be effective in adults who develop this condition [56,57]. The greater benefit of probiotics in this condition is likely because the inflammation occurring in this process is different than the autoimmune-mediated process occurring with IBD.

Allergic problems in childhood are increasingly problematic [58]. The modern diet is estimated to contain significantly fewer bacteria than earlier diets. The hygiene hypothesis has suggested that the lack of exposure to microbes that stimulate the immune system in early infancy is one explanation for the increasing prevalence of allergies and asthma [59]. In children, allergic disorders most often are manifested by skin disruptions, particularly eczema, and respiratory problems including rhinitis and asthma.

Immune system dysregulation appears to be the cause; however, the exact mechanisms are understood incompletely. Allergic immune responses in people are caused by an abnormal T-helper type 2 (Th-2) immune response. The Th-2 stimulation produces proinflammatory substances such as interleukin (IL)-4 and -5, and stimulation of IgE and eosinophil production. The type of bacterial exposure immediately after birth appears to be the prime regulator of the immune system focus toward Th-1 or Th-2. Bacterial exposure commonly seen in breastfed infants creates a focus toward Th-1 responses thereby being less likely to elicit allergic reactions. More recently, it has been shown that reduced activity of T-regulator cells created by flora common in formula-fed infants may contribute to the development of allergic disorders [60].

Numerous in vitro studies have been performed to more clearly elucidate the effect of various probiotics on the immune system [61–63]. In general, these studies have focused on the ability of the bacteria to prevent the Th-2 immune responses to allergens and redirect the system to a Th-1 focus. A recent study has demonstrated that the combination of certain probiotic strains with allergens was effective in creating an effective mucosal vaccination [64].

Bjorksten and colleagues demonstrated a difference in bacterial colonization between children with and without allergies [65]. Two-year-old children with allergies were found to have fewer lactobacilli than nonallergic children, and those without allergies were found to have higher numbers of coliforms and *Staphylococcus aureus*. More recently, it has been demonstrated that infants with allergic symptoms have colonization with bifidobacteria species more commonly found in adults [66].

Atopic eczema associated with cow's milk allergy in infancy was evaluated by Majamaa and Isolauri using a hydrolyzed formula with and without LGG [67]. Those infants receiving the probiotic formula showed a significant improvement in the clinical score of eczema and measurable changes in markers of intestinal inflammation. Mechanistically, it has been shown that LGG, when given with an elimination diet to children with atopic dermatitis caused by presumed cow's milk allergy, decreases IL-4 and increases interferon, redirecting the immune system to decrease inflammation [68]. A study in older children with eczema using *L rhamnous* 19070-2 and *L reuteri* DSM also demonstrated a positive clinical impact of the combination, especially in the group with a positive skin prick test and high serum IgE level [69]. In a recent study, however, infants with cow's milk allergy were given either a single probiotic LGG, a four-strain mixture including LGG, or a placebo, and only the LGG group demonstrated significant changes in the measured variables [70]. This indicates that combination products should be studied as unique substances and should not be presumed to have the properties of the individual components.

Prevention of allergy in infants with a family history of atopy also has been studied. LGG was administered to a group of pregnant mothers and their infants up to 6 months of age who had a family history of atopy and demonstrated a significant difference in the development of eczema in the treatment group at 2 years of age [71]. This group was followed longitudinally, and at 4 years of age, the positive protective effect of the probiotic use remained present [72]. Schultz and colleagues demonstrated ongoing colonization with LGG in a small group of vaginally born infants when their mothers received LGG during the last weeks of pregnancy [73].

Other uses in children

Necrotizing enterocolitis (NEC) is a condition primarily effecting premature infants, and it is thought to be caused by multiple factors including altered enteric mucosa and possible pathogenic organisms. A combination probiotic including *L acidophilus* and *B infantis* was given to neonates in an ICU setting and demonstrated a reduction in NEC [74]. Dani and colleagues, however, were unable to demonstrate any benefit using LGG for preventing urinary tract infection, bacterial sepsis, or NEC in a double-blind study in neonates [75]. A recent study by Lin and colleagues demonstrated a decline in the incidence and severity of NEC in very low birth weight infants when a combination of *L acidophilus* and *B infantis* was given with breast milk, compared with a group fed breastmilk alone [76].

Prevention of systemic infections using probiotics has been suggested by studies showing an enhanced immune response to vaccines [77,78]. A recent study by Hatakka and colleagues suggested that the prolonged use of LGG reduced the incidence of infections in a child care setting [79]. Previously, a study also had shown that the use of LGG decreased the frequency of antibiotic requirements in a group of children with cystic fibrosis [80].

Safety

Probiotic administration to infants and children must be evaluated thoroughly to justify routine usage. Several studies have demonstrated that the oral consumption of lactobacilli and bifidobacteria did not increase the risk for disease even in immunocompromised patients [81,82]. There have been reports of *Lactobacillus* septicemia in children receiving LGG, but all were inpatients and had indwelling central venous catheters [83,84]. In view of the overwhelming association of central venous catheters and probiotic-induced sepsis, central line contamination should be avoided when caring for patients receiving probiotics. As all bacteria are potentially opportunistic, care must be taken to understand the characteristics of any probiotic administered to infants and children. Some lactic acid bacteria produce D-lactate in addition to L-lactate, D-lactate is metabolized poorly in some children and can be associated with various neurological symptoms [85].

SUMMARY

Recently the European Society of Pediatric Gastroenterology, Hepatology and Nutrition Committee on Nutrition issued a position paper on probiotic use in infants [86]. The committee recognized the validity of current studies demonstrating benefits in the health and well being of the infants, particularly in the area of allergic disorders. The committee, however, recognizes that further studies on the safety of the use in infants, particularly those with compromised immune systems, are needed. Formulas with added probiotics appear to have no negative impact on growth and development [87].

Several criteria have been suggested to establish a bacterial strain as a true probiotic. Bacteria frequently identified as of human origin, as having viability in the GI tract, with measurable physiological effects, scientifically documented beneficial effects, and stability during production and packaging are the key elements of a good probiotic. These properties should be well documented, and each species/strain should not be related to another without validated proof. These criteria have established a baseline of requirements, but ongoing research in this area will lead to further refinements. In addition to the benefits derived from modification of newborn intestinal flora, the long-term implications of probiotic use to the healthy child need to be elucidated.

References

[1] Metchnikoff E. The prolongation of life. New York: Putnam & Sons; 1908.
[2] Vergin F. Antibiotics and probiotics. Hippokrates 1954;25(4):116–9.
[3] Fuller R. Probiotics in man and animals. J Appl Bacteriol 1989;66(5):365–78.
[4] Mack DR, Michail S, Wei S, et al. Probiotics inhibit enteropathogenic *E coli* adherence in vitro by inducing intestinal mucin gene expression. Am J Physiol 1999;276:G941–50.
[5] Cebra JJ. Influences of microbiota on intestinal immune system development. Am J Clin Nutr 1999;69(5):1046S–51S.
[6] Edwards CA, Parrett AM. Intestinal flora during the first months of life: new perspectives. Br J Nutr 2002;88(Suppl 1):S11–8.
[7] Mackie RI, Sghir A, Gaskins HR. Developmental microbial ecology of the neonatal gastrointestinal tract. Am J Clin Nutr 1999;69(5):1035S–45S.
[8] Gibson GR, Roberfroid MB. Dietary modulation of the human colonic microbiota: introducing the concept of prebiotics. J Nutr 1995;125(6):1401–12.
[9] Knol J, Schalten P, Kafka C, et al. Colon microflora in infants fed formula with galacto- and fructo-oligosaccharides: more like breastfed infants. J Pediatr Gastroenterol Nutr 2005;40(1):36–42.
[10] Ten Bruggencate SJ, Boree-Oudenhoven JM, Lettink-Wissenk ML, et al. Dietary fructo-oligosaccharides and inulin decrease resistance of rats to *Salmonella*: protective role of calcium. Gut 2004;53(4):530–5.
[11] Harmsen HJ, Wildeboer-Veloo AC, Raango GC, et al. Analysis of intestinal flora development in breast-fed and formula-fed infants by using molecular identification and detection methods. J Pediatr Gastroenterol Nutr 2000;30(1):61–7.
[12] Fuller R. Probiotics in human medicine. Gut 1991;32(4):439–42.
[13] Kalliomaki MA, Isolauri E. Probiotics and down-regulation of the allergic response. Immunol Allergy Clin North Am 2004;24(4):739–52 [viii.].
[14] Lien EL, Davis AM, Euler AR, et al. Growth and safety in term infants fed reduced-protein formula with added bovine alpha-lactalbumin. Nutritional and physiologic significance of alpha-lactalbumin in infants. Bovine chromosome 4 workshop: consensus and comprehensive linkage maps. J Pediatr Gastroenterol Nutr 2004;38(2):170–6.
[15] Bode L, Beemann C, Mank M, et al. Human and bovine milk gangliosides differ in their fatty acid composition. J Nutr 2004;134(11):3016–20.
[16] Midtvedt AC, Midtvedt T. Production of short chain fatty acids by the intestinal microflora during the first 2 years of human life. J Pediatr Gastroenterol Nutr 1992;15(4):395–403.
[17] Edwards CA, Parreh AM, Balmer SE, et al. Faecal short chain fatty acids in breast-fed and formula-fed babies. Acta Paediatr 1994;83(5):459–62.
[18] Parrett AM, Edwards CA. In vitro fermentation of carbohydrate by breastfed and formula-fed infants. Arch Dis Child 1997;76(3):249–53.
[19] Parrett AM, Edwards CA, Lokerse E. Colonic fermentation capacity in vitro: development during weaning in breastfed infants is slower for complex carbohydrates than for sugars. Am J Clin Nutr 1997;65(4):927–33.
[20] Stark PL, Lee A. The microbial ecology of the large bowel of breast-fed and formula-fed infants during the first year of life. J Med Microbiol 1982;15(2):189–203.

[21] Bennet R, Eriksson M, Nord CE, et al. Fecal bacterial microflora of newborn infants during intensive care management and treatment with five antibiotic regimens. Pediatr Infect Dis 1986;5(5):533–9.

[22] Gibson GR, Roberfroid MB, editors. Colonic microbiota, nutrition and health. 1st edition. Kluwer Academic Publishers: Dodrecht; 1999.

[23] Riordan SM, McIver CJ, Wakefield D, et al. Small intestinal mucosal immunity and morphometry in luminal overgrowth of indigenous gut flora. Am J Gastroenterol 2001;96(2):494–500.

[24] Govan JR. In vivo significance of bacteriocins and bacteriocin receptors. Scand J Infect Dis Suppl 1986;49:31–7.

[25] Marteau P, Shanahan F. Basic aspects and pharmacology of probiotics: an overview of pharmacokinetics, mechanisms of action and side effects. Best Pract Res Clin Gastroenterol 2003;17(5):725–40.

[26] Wollowski I, Rechkemmer G, Pool-Zobel BL. Protective role of probiotics and prebiotics in colon cancer. Am J Clin Nutr 2001;73(Suppl 2):451S–5S.

[27] Szajewska H, Mrukowicz JZ. Probiotics in the treatment and prevention of acute infectious diarrhea in infants and children: a systematic review of published randomized, double-blind, placebo-controlled trials. J Pediatr Gastroenterol Nutr 2001;33(Suppl 2):S17–25.

[28] Allen SJ, Okoko B, Martinez E, et al. Probiotics for treating infectious diarrhoea. Cochrane Database Syst Rev 2004;2:CD003048.

[29] Isolauri E, Juntunen M, Rautanen T, et al. A human *Lactobacillus* strain (*Lactobacillus casei* sp strain GG) promotes recovery from acute diarrhea in children. Pediatrics 1991;88(1): 90–7.

[30] Raza S, Graham SM, Allen SJ, et al. *Lactobacillus* GG promotes recovery from acute nonbloody diarrhea in Pakistan. Pediatr Infect Dis J 1995;14(2):107–11.

[31] Shornikova AV, Isolaur E, Burkanova L, et al. A trial in the Karelian Republic of oral rehydration and *Lactobacillus* GG for treatment of acute diarrhoea. Acta Paediatr 1997;86(5):460–5.

[32] Pant AR, Graham SM, Allen SJ, et al. *Lactobacillus* GG and acute diarrhoea in young children in the tropics. J Trop Pediatr 1996;42(3):162–5.

[33] Rautanen T, Isolauri E, Salo E, et al. Management of acute diarrhoea with low osmolarity oral rehydration solutions and *Lactobacillus* strain GG. Arch Dis Child 1998;79(2):157–60.

[34] Shornikova AV, Casas IA, Isolauri E, et al. *Lactobacillus reuteri* as a therapeutic agent in acute diarrhea in young children. J Pediatr Gastroenterol Nutr 1997;24(4):399–404.

[35] Majamaa H, Isolauri E, Saxelin M, et al. Lactic acid bacteria in the treatment of acute rotavirus gastroenteritis. J Pediatr Gastroenterol Nutr 1995;20(3):333–8.

[36] Costa-Ribeiro H, Riboro TC, Mattos AP, et al. Limitations of probiotic therapy in acute, severe dehydrating diarrhea. J Pediatr Gastroenterol Nutr 2003;36(1):112–5.

[37] Salazar-Lindo E, Miranda-Langschwager P, Campos-Sanchez M, et al. *Lactobacillus casei* strain GG in the treatment of infants with acute watery diarrhea: a randomized, double-blind, placebo controlled clinical trial [ISRCTN67363048]. BMC Pediatr 2004;4(1):18.

[38] Pearce JL, Hamilton JR. Controlled trial of orally administered lactobacilli in acute infantile diarrhea. J Pediatr 1974;84(2):261–2.

[39] Isolauri E, Kaila M, Mykkanen H, et al. Oral bacteriotherapy for viral gastroenteritis. Dig Dis Sci 1994;39(12):2595–600.

[40] Oberhelman RA, Gilman RH, Sheen P, et al. A placebo-controlled trial of Lactobacillus GG to prevent diarrhea in undernourished Peruvian children. J Pediatr 1999;134(1):15–20.

[41] Guandalini S, Pensabene L, Zikri MA, et al. *Lactobacillus* GG administered in oral rehydration solution to children with acute diarrhea: a multicenter European trial. J Pediatr Gastroenterol Nutr 2000;30(1):54–60.

[42] Saavedra JM, Bauman NA, Oung I, et al. Feeding of *Bifidobacterium bifidum* and *Streptococcus thermophilus* to infants in hospital for prevention of diarrhoea and shedding of rotavirus. Lancet 1994;344(8929):1046–9.

[43] Thibault H, Aubert-Jacquin C, Goulet O. Effects of long-term consumption of a fermented infant formula (with *Bifidobacterium breve* c50 and *Streptococcus thermophilus* 065) on acute diarrhea in healthy infants. J Pediatr Gastroenterol Nutr 2004;39(2):147–52.
[44] Gianino P, Mastretta E, Longo P, et al. Incidence of nosocomial rotavirus infections, symptomatic and asymptomatic, in breast-fed and non-breastfed infants. J Hosp Infect 2002;50(1):13–7.
[45] D'Souza AL, Raj Kumar C, Cooke J, et al. Probiotics in prevention of antibiotic associated diarrhoea: meta-analysis. BMJ 2002;324(7350):1361.
[46] Tankanow RM, Ross MB, Ertel IJ, et al. A double-blind, placebo-controlled study of the efficacy of Lactinex in the prophylaxis of amoxicillin-induced diarrhea. DICP 1990;24(4):382–4.
[47] Cremonini F, DiCaro S, Covino M, et al. Effect of different probiotic preparations on anti-helicobacter pylori therapy-related side effects: a parallel group, triple blind, placebo-controlled study. Am J Gastroenterol 2002;97(11):2744–9.
[48] Biller JA, Katz AJ, Flores AF, et al. Treatment of recurrent Clostridium difficile colitis with Lactobacillus GG. J Pediatr Gastroenterol Nutr 1995;21(2):224–6.
[49] McFarland LV, Elmer GW, Surawicz CM. Breaking the cycle: treatment strategies for 163 cases of recurrent Clostridium difficile disease. Am J Gastroenterol 2002;97(7):1769–75.
[50] Surawicz CM, McFarland LV, Greenberg RN, et al. The search for a better treatment for recurrent *Clostridium difficile* disease: use of high-dose vancomycin combined with *Saccharomyces boulardii*. Clin Infect Dis 2000;31(4):1012–7.
[51] Schultz M, Veltkamp C, Dieleman LA, et al. Lactobacillus plantarum 299V in the treatment and prevention of spontaneous colitis in interleukin-10-deficient mice. Inflamm Bowel Dis 2002;8(2):71–80.
[52] Dieleman LA, Goerres MS, Arend A, et al. Lactobacillus GG prevents recurrence of colitis in HLA-B27 transgenic rats after antibiotic treatment. Gut 2003;52(3):370–6.
[53] Gupta P, Andrew H, Kirschner BS, et al. Is *Lactobacillus* GG helpful in children with Crohn's disease? Results of a preliminary, open-label study. J Pediatr Gastroenterol Nutr 2000;31(4):453–7.
[54] Group LMS. A Multi-center, placebo-controlled, double blind study of *Lactobacillus GG* in addition to standard maintenance therapy in children with Crohn's disease. J Pediatr Gastroenterol Nutr 2002;35(3):406.
[55] Ishikawa H, Akedo I, Umesaki Y, et al. Randomized controlled trial of the effect of bifidobacteria-fermented milk on ulcerative colitis. J Am Coll Nutr 2003;22(1):56–63.
[56] Gionchetti P, Rizzello F, Venturi A, et al. Oral bacteriotherapy as maintenance treatment in patients with chronic pouchitis: a double-blind, placebo-controlled trial. Gastroenterology 2000;119(2):305–9.
[57] Kuisma J, Mentula S, Jarvinen H, et al. Effect of *Lactobacillus rhamnosus* GG on ileal pouch inflammation and microbial flora. Aliment Pharmacol Ther 2003;17(4):509–15.
[58] Holgate ST. The epidemic of asthma and allergy. J R Soc Med 2004;97(3):103–10.
[59] Rautava S, Ruuskanen O, Ouwehand A, et al. The hygiene hypothesis of atopic disease–an extended version. J Pediatr Gastroenterol Nutr 2004;38(4):378–88.
[60] Romagnani S. The increased prevalence of allergy and the hygiene hypothesis: missing immune deviation, reduced immune suppression, or both? Immunology 2004;112(3):352–63.
[61] Pessi T, Sutas Y, Marttinen A, et al. Probiotics reinforce mucosal degradation of antigens in rats: implications for therapeutic use of probiotics. J Nutr 1998;128(12):2313–8.
[62] Sudo N, Yu XN, Aiba Y, et al. An oral introduction of intestinal bacteria prevents the development of a long-term Th2-skewed immunological memory induced by neonatal antibiotic treatment in mice. Clin Exp Allergy 2002;32(7):1112–6.
[63] Prioult G, Fliss I, Pecquet S. Effect of probiotic bacteria on induction and maintenance of oral tolerance to β-lactoglobulin in gnotobiotic mice. Clin Diagn Lab Immunol 2003;10(5):787–92.

[64] Repa A, Grangette C, Daniel C, et al. Mucosal co-application of lactic acid bacteria and allergen induces counter-regulatory immune responses in a murine model of birch pollen allergy. Vaccine 2003;22(1):87–95.
[65] Bjorksten B, Naaber P, Sepp E, et al. The intestinal microflora in allergic Estonian and Swedish 2-year-old children. Clin Exp Allergy 1999;29(3):342–6.
[66] Ouwehand AC, Isolauri E, He F, et al. Differences in Bifidobacterium flora composition in allergic and healthy infants. J Allergy Clin Immunol 2001;108(1):144–5.
[67] Majamaa H, Isolauri E. Probiotics: a novel approach in the management of food allergy. J Allergy Clin Immunol 1997;99(2):179–85.
[68] Pessi T, Sutas Y, Hume M, et al. Interleukin-10 generation in atopic children following oral *Lactobacillus rhamnosus* GG. Clin Exp Allergy 2000;30(12):1804–8.
[69] Rosenfeldt V, Benefeldt E, Nielsen SD, et al. Effect of probiotic Lactobacillus strains in children with atopic dermatitis. J Allergy Clin Immunol 2003;111(2):389–95.
[70] Pohjavuori E, Viljanen M, Korpela R, et al. *Lactobacillus* GG effect in increasing IFN-γ production in infants with cow's milk allergy. J Allergy Clin Immunol 2004;114(1):131–6.
[71] Kalliomaki M, Saliminen S, Arvilommi H, et al. Probiotics in primary prevention of atopic disease: a randomised placebo-controlled trial. Lancet 2001;357(9262):1076–9.
[72] Kalliomaki M, Saliminen S, Poussa T, et al. Probiotics and prevention of atopic disease: 4-year follow-up of a randomised placebo-controlled trial. Lancet 2003;361(9372): 1869–71.
[73] Schultz M, Gottl C, Young RJ, et al. Administration of oral probiotic bacteria to pregnant women causes temporary infantile colonization. J Pediatr Gastroenterol Nutr 2004; 38(3):293–7.
[74] Hoyos AB. Reduced incidence of necrotizing enterocolitis associated with enteral administration of Lactobacillus acidophilus and Bifidobacterium infantis to neonates in an intensive care unit. Int J Infect Dis 1999;3(4):197–202.
[75] Dani C, Biadaioli R, Bertini G, et al. Probiotics feeding in prevention of urinary tract infection, bacterial sepsis and necrotizing enterocolitis in preterm infants. A prospective double-blind study. Biol Neonate 2002;82(2):103–8.
[76] Lin HC, Su BH, Chen AC, et al. Oral probiotics reduce the incidence and severity of necrotizing enterocolitis in very low birth weight infants. Outcome and cost of intensive care for very low birth weight infants. Pediatrics 2005;115(1):1–4.
[77] Jung L. *Lactobacillus* GG augments the immune response to typhoid vaccination: a double-blinded, placebo-controlled study. FASEB J 1999;13:A872.
[78] Isolauri E, Joensuu J, Suomalainen H, et al. Improved immunogenicity of oral D $\times$ RRV reassortant rotavirus vaccine by *Lactobacillus casei* GG. Vaccine 1995;13(3):310–2.
[79] Hatakka K, Savilahti E, Ponka A, et al. Effect of long-term consumption of probiotic milk on infections in children attending day care centres: double blind, randomised trial. BMJ 2001;322(7298):1327.
[80] Guarino A. Effects of probiotics in children with cystic fibrosis. Gastroenterology International 1998;11:11.
[81] Salminen MK, Rautelin H, Tynkkyen S, et al. *Lactobacillus bacteremia*, clinical significance, and patient outcome, with special focus on probiotic *L rhamnosus* GG. Clin Infect Dis 2004;38(1):62–9.
[82] Salminen MK, Tynkkyen S, Rautelin H, et al. The efficacy and safety of probiotic *Lactobacillus rhamnosus* GG on prolonged, noninfectious diarrhea in HIV patients on antiretroviral therapy: a randomized, placebo-controlled, crossover study. HIV Clin Trials 2004;5(4):183–91.
[83] Land MH, Rouster-Stevens K, Woods CR, et al. Lactobacillus sepsis associated with probiotic therapy. Pediatrics 2005;115(1):178–81.
[84] Kunz AN, Noel JM, Fairchok MP. Two cases of Lactobacillus bacteremia during probiotic treatment of short gut syndrome. J Pediatr Gastroenterol Nutr 2004;38(4):457–8.

[85] Mack DR. D(-)-lactic acid-producing probiotics, D(-)-lactic acidosis and infants. Can J Gastroenterol 2004;18(11):671–5.
[86] Agostoni C, Axelsson I, Braegger C, et al. Probiotic bacteria in dietetic products for infants: a commentary by the ESPGHAN Committee on Nutrition. J Pediatr Gastroenterol Nutr 2004;38(4):365–74.
[87] Saavedra JM, Abi-Hanna A, Moore N, et al. Long-term consumption of infant formulas containing live probiotic bacteria: tolerance and safety. Am J Clin Nutr 2004;79(2):261–7.

Gastroenterol Clin N Am 34 (2005) 465–482

GASTROENTEROLOGY CLINICS
OF NORTH AMERICA

The Role of Enteric Microflora in Inflammatory Bowel Disease: Human and Animal Studies with Probiotics and Prebiotics

Kevin P. Rioux, PhD, MD, FRCPC, Karen L. Madsen, PhD, Richard N. Fedorak, MD, FRCPC*

Department of Medicine, Division of Gastroenterology, University of Alberta, Edmonton, Alberta T6G 2C8, Canada

Bacteria are key initiating factors in the development of chronic intestinal inflammation that characterizes the idiopathic inflammatory bowel diseases (IBDs), ulcerative colitis (UC) and Crohn's Disease (CD). Rather than a single infectious pathogen, the current paradigm holds that IBD is triggered by an abnormal immunological response to commensal bacteria that reside in the gut. It is becoming clear that programmed responses of the immune system are determined genetically and subsequently fine-tuned by environmental factors such as the enteric microbiota. Overall, in IBD the balance between aggressive and beneficial bacterial species appears to favor the development of chronic intestinal inflammation in those genetically predisposed.

In the past 5 years, probiotics and prebiotics have attracted considerable attention from the scientific and medical communities. Probiotics are live microorganisms, which, when consumed in adequate amounts, confer a health benefit to the host. Prebiotics are nondigestible substances that act as a selective nutritive substrate for beneficial colonic bacteria and thereby promote their growth and metabolism to ultimately enhance the health of the host. Once relegated as questionable therapies for an inordinate number of ailments within

Dr. Rioux's research is supported by the Canadian Association of Gastroenterology, Canadian Institutes for Health Research, Crohn's and Colitis Foundation of Canada, Alberta Heritage Foundation for Medical Research. Dr. Madsen's research is supported by the Canadian Institutes for Health Research, Crohn's and Colitis Foundation of Canada, Alberta Heritage Foundation for Medical Research. Dr. Fedorak's research is supported by the Canadian Institutes for Health Research, Crohn's and Colitis Foundation of Canada.

*Corresponding author. Division of Gastroenterology, Zeidler Family Gastrointestinal Health and Research Centre, University of Alberta, Suite 205 College Plaza, 8215–112 Street Edmonton, Alberta T6G 2C8, Canada. *E-mail address*: richard.fedorak@ualberta.ca (R.N. Fedorak).

0889-8553/05/$ – see front matter

doi:10.1016/j.gtc.2005.05.005
gastro.theclinics.com

the vast realm of complementary medicine, scientific data are emerging about the biological effects of these agents, and important therapeutic applications seem eminent in the treatment of allergic, autoimmune, and inflammatory disease. By altering the enteric microbiota, either through ingestion of prebiotic or probiotic agents, it may be possible to ameliorate inflammation and prevent its recurrence.

This article reviews accumulated evidence implicating commensal bacteria as trigger and perpetuator of gut inflammation. The biological effects of probiotics and prebiotics are described, serving to highlight their therapeutic potential. Efficacy data from studies in which prebiotics and probiotics have been tested in experimental colitis as well as human IBD are reviewed.

ENTERIC BACTERIA: A KEY INITIATOR OF INFLAMMATORY BOWEL DISEASE

Because of similarities between infectious enterocolitis and idiopathic IBD, there has been a strong hope and suspicion that IBD may be caused simply by a single infectious agent. Despite reasonable rationale and effort, purported links, for example between measles virus or *Mycobacterium paratuberculosis* and IBD, have not led to consistent or conclusive findings. Although multiple lines of evidence point against a single microbial cause of IBD [1], the complexity of the enteric microflora, the relatively unstudied realm of enteric retroviruses, and numerous surprising lessons from medical history suggest that a single infectious agent may still underlie the development of IBD in a subset of patients. However, the prevailing concept is that nonpathogenic bacteria traditionally regarded as commensal organisms may, in fact, be the essential cause of IBD in the presence of a permissive immunogenetic background.

Several experimental models of IBD have been developed and, in each model, sporadic or engineered mutations in key controlling elements of the immune system result in spontaneous development of chronic intestinal inflammation reminiscent of IBD. Of importance, and almost invariably, intestinal inflammation does not develop under germ-free conditions. Two important conclusions therefore emerge: (1) nonpathogenic bacteria can evoke chronic intestinal inflammation if certain genetically determined immunologic susceptibilities exist and (2) a variety of immunological imbalances can predispose to a common indistinguishable outcome–chronic intestinal inflammation.

Clues to the essential etiologic role of bacteria in inflammatory bowel disease are summarized in Box 1. A simple observation is that IBD tends to occur in the areas of the intestine that harbor the highest concentrations of luminal and mucosal bacteria. This is the case for ulcerative colitis, where inflammation invariably involves the rectum, where bacterial numbers in stool approach 10^{11} to 10^{12} per gram, and contact time is greatest owing to its function as a reservoir. Bacterial counts in the ileum, by contrast, are 1000- to 10,000-fold lower than in the colon. The common and often exclusive involvement of the

Box 1: Evidence of essential etiologic role of bacteria in inflammatory bowel disease

- The enteric microbiota is a dynamic influence on mucosal immune development and function.
- Inflammation is greatest in the area with highest concentrations of luminal bacteria.
- Dysbiosis has been observed among patients who have IBD.
- Antibiotics ameliorate intestinal inflammation.
- Surgical diversion of fecal stream effectively abolishes recurrence of IBD; restoration of intestinal continuity precipitates anastomotic recurrence.
- Enteroadherent and intramucosal bacteria are increased in Crohn's disease.
- NOD2/CARD15 polymorphisms associated with CD result in aberrant mucosal immune defenses against bacteria.
- Loss of immunologic tolerance to commensal gut bacteria is demonstrated in IBD.
- Animals that are genetically predisposed to intestinal inflammation do not develop this outcome in the absence of luminal bacteria.
- In animal models of colitis, disease is evoked after adoptive transfer of T cells with specific reactivity to enteric bacteria, but not dietary or epithelial antigens.

ileum in CD may reflect the robust mucosal immune system at this site in combination with a formidable bacterial load and contact time prolonged by the ileocecal valve. In CD, surgical diversion of the fecal stream is associated with disease improvement, and disease recurs when continuity of the bowel is restored or when luminal contents containing bacteria are reinfused directly into the bypassed segment [2,3]. Ultrafiltrate of luminal contents devoid of bacteria do not reproduce this finding [2].

The community of bacteria in direct contact with the intestinal mucosa is arguably the microniche that is most relevant to IBD. Swidsinski and colleagues demonstrated that patients who have IBD have a greater number of mucosa-associated bacteria, particularly *Bacteroides* species and *Escherichia coli* [4]. This does not appear to be an artifact of intestinal inflammation or epithelial disruption, as the high density of mucosa-associated bacteria in patients who have CD also occurred in samples of noninflamed intestine, suggesting that altered bacterial epithelial adherence may be a cause rather than a consequence of the disease. Indeed, strains of *E coli* with aggressive adherent–invasive properties have been isolated from patients with CD and associated with the propensity for postoperative recurrence [5,6]. In patients who have IBD, there appears to be a genetically determined loss of immunologic tolerance, as evidenced by specific cell-mediated and humoral immune responses targeting commensal gut bacteria [7,8].

Originally a concept advanced by Metchnikoff in the early 1900s [9], dysbiosis refers to a maladaptive alteration of the intestinal microbiota, shifting the population balance in favor of putative harmful bacteria. The term has been applied in the setting of IBD to describe various changes in the number and metabolism of intestinal bacteria in patients with UC, CD, and pouchitis, as has been reviewed recently [10]. In general, increased numbers of *Bacteroides* species and members of the genus *Enterobacteriaceae* have been described in IBD and associated with decreased numbers of presumably beneficial bacteria such as bifidobacteria and lactobacilli. Although definitive proof is lacking, it has been suggested that dysbiosis may exist as a necessary precondition to intestinal inflammation. Genetic and environmental factors may underlie dysbiosis in IBD.

Exposure of the intestine to commensal bacteria early in life appears to be vital to the optimal education and development of the mucosal and systemic immune systems. Through competitive interactions for essential nutrients and cross-talk mediated by physical and chemical means (quorum-sensing), the intestinal microbiota form a complex ecosystem that is relatively stable in terms of composition and metabolic activity. Similarly dynamic interactions occur between intestinal bacteria and the mucosa that alters local physiology and immunology. These mechanisms are being defined at the molecular level. Thus, early and ongoing interplay between host and autochthonous bacteria likely play a role in chronic intestinal inflammation.

These arguments establish a solid rationale for using antibiotics as primary or adjunctive treatment for IBD. Indeed, antibiotics have been employed successfully for treating pouchitis and septic and perforating complications of IBD, but clear evidence of clinical benefit in luminal IBD disease is somewhat lacking. As this topic is beyond the intended scope of this article, the reader is referred to a recent comprehensive review [11]. The disappointing results of antibiotics in IBD does not necessarily speak against the etiologic role of bacteria, but perhaps reflects only methodologic shortfalls regarding choice of antibiotic agent(s) and spectrum of antimicrobial activity, dose, duration, bioavailability, and phenotype of IBD.

A significant association exists between certain polymorphisms of NOD2/CARD15 gene and a subset of patients who have CD [12,13]. Aberrant NOD2/CARD15 protein has been shown to result in defects in recognition and clearance of invasive bacteria by macrophages, and decreased production of antibacterial defensins by Panneth cells [14–16]. These findings again emphasize the importance of bacteria as a trigger of IBD. The final and perhaps most compelling evidence of the interaction between bacteria, immunity, and genes comes from a series of sporadic, engineered, and genetically determined animal models of IBD. In each of these models, enteric bacteria are required for the induction of the inflammatory process, regardless of the underlying immunological defect. Moreover, mono-association experiments have demonstrated that individual commensal bacteria vary widely in their ability to evoke intestinal inflammation in genetically predisposed hosts. These models have served as a useful testing ground for the effectiveness of various probiotic and prebiotic therapies, as will be discussed in the next sections.

MECHANISM OF ACTION OF PROBIOTICS: ANIMAL AND IN VITRO MODELS

Experiments performed in animal and in vitro models have helped to define an array of biological effects that probiotic organisms have. The beneficial effects of probiotic bacteria can be classified into three main categories: (1) alteration of the enteric microbiota, (2) modulation of the host immune response, and (3) enhancement of barrier function through interactions with epithelial and immune cells in the gut. These principal effects are outlined in Box 2, and a few salient examples are described. Further details are available in a recent review [17].

Alteration of the enteric microbiota

Probiotic organisms transiently reside in the gut and have biological impact simply by competing for nutrients and space, a concept akin to colonization resistance within the indigenous community of gut bacteria. By secreting short chain organic acids and peptides with defined bacteriocidal activities, probiotic organisms may render the luminal environment unfavorable for more aggressive

Box 2: Biological effects of probiotic bacteria demonstrated in experimental models

Enhanced intestinal epithelial barrier function

- Enhance transepithelial resistance
- Enhance mucus production
- Modify epithelial glycosylation
- Bolster epithelial cytoskeletal and tight junctional integrity
- Promote epithelial restitution
- Prevent epithelial cell apoptosis
- Antioxidative effects

Modulation of the immune system

- Increase total and specific IgA production
- Down-regulate T-cell responsiveness
- Decrease phagocytic activity
- Enhance apoptosis of immunocytes
- Alter cytokine profiles
- Induce oral tolerance

Direct alteration of the intestinal microbiota

- Interfere with pathogen adherence and invasion
- Bacteriocin production
- Decrease luminal pH by means of formation of organic acids
- Colonization resistance

bacteria. For example, in a coculture system and a gnotobiotic mouse model, the probiotic organism *Bifidobacterium infantis* suppresses growth of *Bacteroides vulgatus,* which has a putative role in IBD [18]. In this manner, probiotics may restore a more favorable balance between beneficial and detrimental bacteria, correcting the dysbiosis that has been described in IBD patients.

Enhance barrier function

By up-regulating mucin secretion [19], or occupying bacterial receptors on the intestinal epithelium or altering their glycosylation [20], probiotics may prevent pathogen adherence and invasion. *E coli* Nissle 1917, for example, protects against an adherent-invasive strain of *E coli* isolated from patients who have CD [21]. Madsen and colleagues have demonstrated that oral administration of VSL#3, a mixture of eight different probiotic organisms (*Lactobacillus casei, L plantarum, L acidophilus, and L delbrueckii* subspecies bulgaricus*; B longum, B breve, and B infantis;* and *Streptococcus salivarius* subspecies thermophilus), enhances colonic epithelial integrity and attenuates the development of colitis in interleukin (IL)-10 gene-deficient mice [22]. Nonviable components appear to mediate this effect, as a soluble factor derived from VSL#3-conditioned media could enhance barrier function in epithelial cell monolayers. Viable *S thermophilus* and *L acidophilus* appear to maintain and even bolster epithelial integrity by direct actions on epithelial tight junction and cytoskeletal elements [23].

Local and systemic immunomodulation

Probiotic organisms have been shown to favorably influence antigen presentation, humoral- and cell-mediated immune responses, and induce intestinal production of anti-inflammatory cytokines (eg, IL-10, transforming growth factor-β) while down-regulating production of proinflammatory cytokines (eg, tumor necrosis factor α [TNF-α], IL-8) [17]. Recent studies have shown that nonmethylated bacterial DNA (CpG) from some VSL#3 components may mediate the beneficial effects observed in experimental colitis by inhibiting epithelial proinflammatory responses and apoptosis [24,25]. Probiotic-derived CpG [24,25] and living probiotic organisms [26] appear to ameliorate experimental colitis when administered parenterally also. These recent findings challenge the notion that the beneficial effects of probiotics relate to interaction of living probiotic organisms with the intestinal epithelium and resident commensals.

In addition to the mechanistic studies, animal models of IBD have been valuable in evaluating the therapeutic effectiveness of probiotics. This has been studied primarily in the IL-10 gene-deficient mouse and the HLA-B27 transgenic rat.

Interleukin-10 gene-deficient (IL-10 −/−) mice spontaneously develop intestinal inflammation with numerous features similar to Crohn's colitis. These animals have an altered complement of intestinal commensal bacteria, with increased mucosa-associated aerobic species and decreased lactobacilli, which provided a solid rationale for early studies of probiotics in this model. In IL-10 −/− mice, colitis is attenuated significantly by enteral administration of various strains of both lactobacilli and bifidobacteria [22,27–30]. Although VSL#3 [22] and *L plantarum* 299v [28] have been shown to be effective primary treatment for established colitis in the IL-10 –/– mouse model, the most consistent efficacy of appears to be in preventing onset of inflammation before it spontaneously occurs. This concept is reinforced by studies in the HLA-B27 model of spontaneous colitis, where *L casei* subspecies *rhamnosus* (*LGG*) effectively prevented development of colitis but could not ameliorate established disease [31].

The benefit of a particular probiotic is not universal across genetic platforms, as demonstrated by the fact that *L plantarum* 299v ameliorates colitis in the IL-10 –/– gene-deficient mouse, but not in HLA-B27 rat model. This finding holds some caution about the promise of being able to develop a single probiotic that is universally effective, considering the putative genetic and bacteriological complexity of human IBD.

Probiotics have not been particularly successful in treating experimental colitis induced by hapten trinitrobenzene sulfonic (TNBS) acid [32,33]. Similarly, in the dextran sulfate sodium (DSS) model of colitis, *E coli* Nissle 1917 did not reduce the histologic severity of inflammation, despite having a favorable effect on some mucosal proinflammatory cytokines [34]. In contrast to their ineffectiveness in TNBS colitis, both VSL#3 and *LGG* ameliorate colitis induced by the sulfhydryl blocker iodoacetamide. With a clear role of genetic susceptibilities in IBD and ample description of immunomodulatory effects of probiotics, the mechanistic implications of studying probiotics in chemically-induced colitis seem questionable.

In general, significant differences exist among various probiotic bacterial species and strains in terms of biological effect. Indeed, multiple mechanisms may be exploited in the development of optimal probiotic therapies. The convincing efficacy of VSL#3 in various experimental models of colitis and in treating human pouchitis, for example, underscores the fact that success with probiotics may require the use of multiple organisms, each exploiting one or more biological effects. A greater understanding of these mechanisms of action and high throughput methodologies for testing candidate probiotics will facilitate development of probiotic strains for specific clinical indications. It also will allow for the determination of optimal doses, timing of administration, and potential synergy between probiotic species.

Considering that the effects of probiotic bacteria on the host may be founded in genetically and developmentally programmed immune responses, it may be difficult to predict how an individual patient will respond to a particular probiotic. This only adds one more level of complexity to the current challenges of applying probiotics to human disease. The real utility of probiotics may lie in their ability to act as carriers for the precise delivery of topical biological agents. In this regard, *L lactis* has been modified genetically to produce and release high concentrations of the anti-inflammatory cytokine IL-10 in the gut lumen [35], thereby ameliorating intestinal inflammation in the IL-10 gene-deficient mouse model of enterocolitis [36].

POUCHITIS

Pouchitis is a nonspecific inflammation of the ileal reservoir and is a common complication following ileal pouch anal anastomosis (IPAA) surgery for ulcerative colitis. Its clinical frequency depends largely on the duration of the follow-up, but is approximately 50% after 10 years. The cause of pouchitis remains largely unknown, but some studies have demonstrated reduced counts of lactobacilli and bifidobacteria within the pouch [37], suggesting that dysbiosis may contribute to the development of this syndrome. Although most patients with pouchitis respond to treatment with antimicrobial agents, approximately 10% of patients experience recurrent or refractory disease. It was through well-designed clinical trials in patients with pouchitis that the therapeutic potential of probiotics for IBD initially was demonstrated (Table 1).

In a landmark trial of probiotics in IBD, Gionchetti and colleagues [38] studied the secondary prevention of pouchitis in 40 patients in whom acute pouchitis had been treated successfully with antibiotics. Once in remission, the patients were randomized to orally receive VSL#3 or placebo and studied prospectively for 9 months. Of the 20 patients who received placebo, 100% had relapses within the study period. In contrast, 17 of the 20 patients (85%) who received VSL#3 remained in remission, with a pouchitis disease activity index of zero (normal), and, moreover, viable probiotic organisms derived from VSL#3 could be recovered from these patients. When VSL#3 therapy was discontinued, all patients developed recurrent pouchitis within 3 months. These findings have been confirmed by others [39], and Gionchetti and colleagues have extended their original findings to show that VSL#3 is effective for primary prevention of pouchitis when oral therapy is begun immediately after IPAA surgery in ulcerative colitis patients [40].

In contrast, Kuisma and colleagues [41] did not demonstrate a therapeutic effect of probiotics in acute pouchitis with *LGG* administered orally for a 3-month period. Only 40% of the study subjects were colonized with *LGG*,

Table 1
Clinical trials of probiotics in pouchitis

Reference	Methodology	N	Treatments	Results
Maintenance of antibiotic-induced remission				
Gionchetti [38]	RCT	40	VSL#3 versus placebo	$P < 0.001$
Mimura [39]	RCT	36	VSL#3 versus placebo	$P < 0.001$
Maintenance of surgically-induced remission				
Gionchetti [40]	RCT	40	VSL#3 versus placebo	$P < 0.05$
Treatment of acute pouchitis				
Kuisma [41]	RCT	20	*Lactobacillus* GG versus placebo	$P = \text{NS}$
Laake [42]	Open-label	10	Cultura	$P = \text{NS}$

which may underlie the study's negative findings. Similarly, no objective benefit was observed when Laake and colleagues [42] treated 10 patients who had acute pouchitis for 4 weeks with Cultura, a fermented dairy product containing *L acidophilus* and *B lactis*.

In summary, the available evidence shows that probiotics are not effective for inducing remission in acute pouchitis, but there is strong evidence to support the use of the probiotic cocktail VSL#3 for primary pouchitis prevention after IPAA surgery and for prevention of pouchitis recurrence after antibiotic-induced remission.

CROHN'S DISEASE

It is unclear whether probiotics are useful for treating active CD, as the quality of data is lacking (Table 2). In a small, double-blind trial, Schultz and colleagues [43] initially treated patients who had CD with antibiotics and a tapering course of glucocorticoids. When the antibiotics were stopped at 2 weeks, they were randomized to receive either *LGG* or placebo. There was no difference in success of achieving remission between the groups (four out of five *LGG* versus five of six placebo), but because of the small sample sizes, no reliable assessment could be made of maintenance of remission at 6 months. In an open-label uncontrolled trial, McCarthy and colleagues showed that oral administration of *L salivarius* UCC118 caused a significant drop in disease

Table 2
Clinical trials of probiotics in Crohn's disease

Reference	Methodology	N	Treatments	Results
Treatment of active Crohn's disease				
Schultz [43]	RCT	11	*Lactobacillus* GG versus placebo (plus initial antibiotics/steroids)	$P = \text{NS}$
McCarthy [44]	Open-label	25	*Lactobacillus salivarius*	↓CDAI ($P < 0.05$)
Gupta [45]	Open-label	4	*Lactobacillus* GG	↓CDAI ($P < 0.02$)
Maintenance of medically induced remission				
Malchow [46]	RCT	28	*E coli* Nissle 1917 versus placebo	$P = \text{NS}$
Guslandi [47]	Open-label	32	*S boulardii* versus mesalamine	$P < 0.05$
Maintenance of surgically induced remission				
Prantera [48]	RCT	45	*Lactobacillus* GG versus placebo	$P = \text{NS}$
Campieri [49]	RCT	40	VSL#3 versus mesalamine	$P < 0.05$

activity in patients who had mild-to-moderately active CD [44]. Gupta and colleagues [45] conducted a 6-month open label trial using *LGG* in four children who had CD in conjunction with concomitant therapy with prednisone and immunomodulatory agents. A significant improvement was noted as early as 1 week after starting therapy, and this was sustained throughout the study and, in three patients, it was possible to taper the dose of steroids. Three patients had relapse of their CD within 4 to 12 weeks of discontinuing the probiotic.

In a randomized, double-blind study, Malchow et al [46] demonstrated that *E coli* Nissle 1917 reduced disease relapse in patients who had active colonic CD and in whom remission had been induced by corticosteroids (70% still in remission in probiotic group at 1 year versus 30% placebo). These results did not achieve statistical significance, however. The only study to demonstrate a statistically significant prolongation of medically induced remission in CD was that of Guslandi and colleagues [47] using *Saccharomyces boulardii* and mesalamine. Clinical remission at 6 months was observed in 10 of the 16 patients on mesalamine maintenance, and in 15 out of 16 patients receiving mesalamine plus *S boulardii*. Despite this encouraging result, no further studies with this probiotic agent have been published.

The benefits of probiotics in surgically induced remission in CD are also not very compelling. In a randomized, double-blind trial, 45 patients received *LGG* or placebo within 10 days following surgical resection of their CD [48]. After 52 weeks of treatment, there was no significant difference in clinical or endoscopic remission between the two groups. Campieri and colleagues [49] reported that a combination of antibiotic and VSL#3 was effective at preventing post-operative recurrence of CD when compared with mesalamine. Forty patients were randomized to receive either rifaxamin for 3 months then followed by VSL#3 for 9 months, or mesalamine for 12 months. After 1 year, the antibiotic/VSL#3 group had an endoscopic remission rate of 80% compared with 60% in the mesalamine group. Interestingly, the endoscopic remission rate at 1 year in the antibiotic/VSL#3 group was similar to that previously described with metronidazole alone [50].

ULCERATIVE COLITIS

Studies of probiotics for treating UC are summarized in Table 3. In the only published randomized, double-blind, placebo-controlled trial examining the effectiveness of probiotics in treating mild-to-moderate UC, Kato and colleagues found that *Bifidobacterium*-fermented milk significantly reduced the clinical activity index at 12 weeks compared with the control group [51]. All patients were taking either sulfasalazine or mesalamine at stable doses continued from before trial entry. Endoscopic scores also improved in the probiotic group but were not significantly different from controls at the 12-week endpoint. A study by Rembacken and colleagues [52] suggested that *E coli* Nissle 1917 was equivalent to mesalamine for maintaining remission in UC. At entry into the study, all patients were given a 1 week course of oral gentamicin to suppress their native *E coli* flora. Along with the study, probiotic or

Table 3
Clinical trials of probiotics in ulcerative colitis

Reference	Methodology	N	Treatments	Results
Treatment of acute ulcerative colitis				
Kato [51]	RCT	20	*Bifidobacterium*-fermented milk versus placebo (both groups on aminosalicylates)	$P < 0.01$
Rembacken [52]	RCT	116	*E coli* Nissle 1917 versus mesalamine (both had prednisolone induction)	$P =$ NS
Fedorak [53]	Open-label	30	VSL#3	63% remission
Guslandi [54]	Open-label	25	*Saccahromyces boulardi*	68% remission
Borody [55]	Open-label	6	Fecal enemas	100% remission
Maintenance of medically-induced remission				
Rembacken [52]	RCT	116	*E coli* Nissle 1917 versus mesalamine	$P =$ NS*
Kruis [56]	RCT	327	*E coli* Nissle 1917 versus mesalamine	Equivalency demonstrated
Ishikawa [57]	RCT	21	Yakult versus placebo	$P < 0.02$
Venturi [58]	Open-label	20	VSL#3	75% still in remission at 1 year

*Study not designed to test equivalency.

mesalamine patients also were permitted to receive treatment with tapering prednisolone for moderate and severe disease, and rectal hydrocortisone enemas for proctitis for 3 months to induce remission. After entering into remission, patients were maintained on either mesalamine or *E coli* Nissle 1917 for up to 12 months. Forty-four patients (75%) in the mesalamine group and 39 patients (68%) in the probiotic group achieved remission. The median time to remission was similar in both groups. At the end of 12 months, 25% of patients in the mesalamine group and 26% of patients in the *E coli* group maintained remission. The median duration of remission was similar in both groups. The authors suggested that the efficacy of treatment with this nonpathogenic *E coli* was equivalent to mesalamine in maintaining remission. Nevertheless, the study was not powered for equivalence, and the remission rates at 1 year were close to historical placebo rates.

Recently, Fedorak and colleagues [53] evaluated the efficacy of VSL#3 in an open-label, two-center trial in 30 patients with mildly to moderately active UC who had failed mesalamine. Remission was achieved in 63% and response in 23% of patients who were treated with VSL#3, representing a notable improvement in a difficult to treat population. A randomized, controlled trial is underway to confirm these results.

Guslandi and colleagues [54] studied 25 patients with a clinical flare of their UC despite being on maintenance doses of mesalamine for at least 3 months. In this open-label trial, patients received *S boulardii* for 4 weeks while mesalamine was continued. A significant reduction in the UC activity index score was observed at the end of treatment, and remission was achieved in 17 patients (68%). The remission rate in this pilot study is remarkably similar to the remission rates described for *E coli* Nissle 1917 and VSL#3.

Borody and colleagues [55] documented favorable responses to bacteriotherapy (fecal enemas) in six patients who previously had failed maximum standard therapy for UC, including multiple courses of corticosteroids and immunosuppressive therapy. In 1 to 13 years of follow-up, there has been no clinical, colonoscopic, or histologic evidence of UC recurrence in any patient, despite the fact that no maintenance therapy has been used. These results are dramatic and worthy of further investigation both from pathogenesis and therapeutic standpoints.

In a double-dummy, double-blind, randomized trial involving 327 patients, Kruis and colleagues [56] studied the effectiveness of an oral preparation of *E coli* Nissle 1917 compared with mesalamine for maintaining remission in UC. At the end of the 12-month study, there was no significant difference between the two study groups, with relapses occurring in 36.4% of the *E coli* Nissle 1917 group and 33.9% of the mesalamine group. This study was powered adequately to demonstrate equivalence, and therefore this particular probiotic organism could be offered as an alternative to mesalamine for maintaining UC remission.

Ishikawa and colleagues [57] conducted a randomized, placebo-controlled trial in 21 patients to examine the effect of fermented milk that contained live bifidobacteria and *L acidophilus* in patients who had UC in remission. At the end of 12 months, clinical remission was seen in 73% of patients taking probiotic versus 10% in the placebo group, but there was no significant difference in colonoscopic findings. A larger, multi-center study with this probiotic product is underway. Finally, in an open-label uncontrolled trial, Venturi and colleagues [58] showed that VSL#3 maintained clinical and endoscopic remission in 75% of the 20 patients who were studied for 1 year.

PREBIOTICS

Prebiotics are nondigestible food ingredients that selectively stimulate growth or metabolic activity of beneficial bacteria in the colon and thereby improve host health [59]. Prebiotic agents include dietary substances such as resistant starch, nonstarch polysaccharides, and various oligosaccharides. For human applications, the greatest interest is in oligosaccharide prebiotics such as fructo–oligosaccharides and inulin.

Prebiotics pass through the upper gastrointestinal tract intact and therefore arrive in the distal gut where they are fermented selectively by indigenous beneficial bacteria. Bifidobacteria and lactobacilli are increased in number following oral administration of certain prebiotics. In human volunteers, fructo-oligosaccharides and inulin increase the number of bifidobacteria and lactobacilli associated with the mucosa by 1 log colony forming unit (CFU)/g mucosa and 0.5 log CFU/g, respectively [60]. Thus, the functional effect of prebiotics is the same as exogenous administration of probiotic lactobacilli and bifidobacteria, and the biological effects of prebiotics are likely the same as probiotics, including enhancement of barrier function and immunomodulatory effects, and favorable alteration of the intestinal microbiota (see Box 2).

The biological and clinical effects of prebiotics have been tested in various animal models. Madsen and colleagues [27] demonstrated that lactulose promotes proliferation of intestinal lactobacilli and was associated with attenuation of colitis in IL-10 gene-deficient mice. Germinated barley foodstuff (GBF), insoluble mixture of glutamine-rich protein and hemicellulose-rich dietary fiber, was shown to significantly attenuate colitis induced by DSS [61]. The therapeutic effect of GBF in this model was accompanied by increased intestinal concentrations of short-chain fatty acids and altered intestinal microbiota, with numerically increased bifidobacteria and eubacteria and diminished numbers of bacteroides and enterobacteriaceae. By similar mechanisms, Cherbut and colleagues showed that fructo-oligosaccharides attenuate TNBS-induced colitis in rats. In HLA-B27 transgenic rats, a combination of inulin and fructo-oligosaccharides was shown to significantly decrease the severity of colitis. Prebiotics are not universally effective in experimental colitis, however, as fructo-oligosaccharides and galacto-oligosaccharide showed no benefit in the DSS and TNBS-colitis models, respectively [62,63].

Prebiotics have been tested to a very limited extent in patients who have UC but, as yet, not in CD. The germinated barley foodstuff initially tested in the DSS experimental colitis model, later was applied to treat patients with mildly to moderately active colitis [64]. In an open-label trial, 21 patients were given 20 to 30g/d of GBF for 24 weeks. At the end of the study, the clinical activity index was improved significantly compared with baseline, and fecal concentrations of bifidobacteria were increased. Recently, GBF was reported to be effective in maintaining remission in UC also [65], and it apparently is tolerated well in this patient population.

There is a lack of high-quality evidence regarding the use of prebiotics in IBD; nonetheless, the available evidence from animal models and patients with colitis underscores the potential of these dietary supplements. Moreover, the combination of probiotics and prebiotics, termed synbiotics, seems a logical progression in this field of study and has met with some early clinical success. In a randomized, placebo-controlled trial, Furrie and colleagues [66] reported the use of *B longum* in combination with a prebiotic for treating patients who had active UC. The prebiotic was inulin and oligo-fructose, which is purported to be a selective substrate to support the growth of the administered bifidobacteria. At the end of the 1-month trial, there was a significant improvement in endoscopic and histologic scores and a decrease in cytokine mediators of inflammation in the synbiotic group.

If efficacy is proven, IBD therapy with prebiotics may be limited by benign but bothersome side effects such as bloating, flatulence, borborygmi, and diarrhea, which may be particularly intolerable in IBD patients. Also of note, there is evidence of chemopreventive properties of prebiotics [67], of added relevance and importance to patients who have UC and Crohn's colitis, given their increased risk for adenocarcinoma of the colon.

SUMMARY

Probiotic and prebiotic agents are capable of altering the enteric microbiota to affect a more favorable balance between beneficial bacteria and those that may be aggressors in conditions like pouchitis and IBD. Some probiotic bacteria appear to have numerous biological effects beyond this, including enhancement of the intestinal epithelial barrier and immunomodulatory functions. Probiotics, prebiotics, and synbiotics have potential as therapeutic agents in gastroenterology, as do probiotic organisms that have been genetically engineered to deliver anti-inflammatory cytokines to the intestinal mucosa.

Despite favorable findings from in vitro studies of probiotics and therapeutic trials in animal models of IBD, early human clinical trials of probiotics have provided mixed results and generally have been of weak methodologic quality. The best evidence for the use of probiotics in IBD comes from studies of the probiotic combination VSL#3 in preventing initial onset and clinical recurrence of pouchitis in patients who have UC and have undergone IPAA. For maintaining remission in UC, there also is evidence that *E coli* Nissle 1917 is as effective as conventional maintenance doses of mesalamine. There is a distinct lack of quality evidence from properly conducted clinical trials supporting the use of probiotics in treating active UC or CD. Overall, animal and human studies suggest that probiotics are therapeutically more useful in preventing recurrence or initial onset of inflammation, but they do not appear to be particularly powerful agents inducing remission. Numerous prebiotics have been tested in experimental models of colitis and have shown quite favorable results, but there are very few clinical trials of prebiotics in people who have IBD.

Much remains unknown about how best to apply probiotics and prebiotics to the treatment of IBD in people. Fundamental questions exist, for example, about which single agent is most effective or whether cocktails of several probiotic bacteria or prebiotics are uniformly superior. Information is lacking about optimal dose, frequency, and duration for either prebiotics or probiotics. No single probiotic bacterium has emerged as therapeutically most promising, and, in fact, the efficacy of a probiotic in treating human IBD may relate to the individual's underlying immunogenetic susceptibilities to IBD and the composition of their intestinal microbiota. Both of these factors are exceedingly complex and varied between individuals with IBD. Therefore, a single agent probiotic may have limited efficacy in the larger population of IBD.

Regardless, to dispel the seemingly endless claims and speculation about the therapeutic benefits of probiotics and prebiotics in IBD, large, randomized, double-blind placebo-controlled clinical trials are needed in a phenotypically well-defined population of IBD patients. In the meantime, basic science will help identify and engineer more probiotic organisms, prebiotic compounds, and combinations, and further elucidate their mechanisms of action.

References

[1] Shanahan F. Probiotics in inflammatory bowel disease—therapeutic rationale and role. Adv Drug Deliv Rev 2004;56:809–18.
[2] Harper PH, Lee EC, Kettlewell MG, et al. Role of the faecal stream in the maintenance of Crohn's colitis. Gut 1985;26(3):279–84.
[3] D'Haens GR, Geboes K, Peeters M, et al. Early lesions of recurrent Crohn's disease caused by infusion of intestinal contents in excluded ileum. Gastroenterology 1998;114(2):262–7.
[4] Swidsinski A, Ladhoff A, Pernthaler A, et al. Mucosal flora in inflammatory bowel disease. Gastroenterology 2002;122(1):44–54.
[5] Darfeuille-Michaud A, Neut C, Barnich N, et al. Presence of adherent *Escherichia coli* strains in ileal mucosa of patients with Crohn's disease. Gastroenterology 1998;115(6): 1405–13.
[6] Darfeuille-Michaud A, Boudeau J, Bulois P, et al. High prevalence of adherent-invasive *Escherichia coli* associated with ileal mucosa in Crohn's disease. Gastroenterology 2004; 127(2):412–21.
[7] Duchmann R, Kaiser I, Hermann E, et al. Tolerance exists towards resident intestinal flora but is broken in active inflammatory bowel disease. Clin Exp Immunol 1995;102(3):448–55.
[8] Macpherson A, Khoo UY, Forgacs I, et al. Mucosal antibodies in inflammatory bowel disease are directed against intestinal bacteria. Gut 1996;38(3):365–75.
[9] Metchnikoff E. The prolongation of life: optimistic studies. London: Heinmann; 1907.
[10] Tamboli CP, Neut C, Desreumaux P, et al. Dysbiosis in inflammatory bowel disease. Gut 2004;53:1–4.
[11] Isaacs KL, Sartor RB. Treatment of inflammatory bowel disease with antibiotics. Gastroenterol Clin North Am 2004;33(2):335–45.
[12] Ogura Y, Bonen DK, Inohara N, et al. A frameshift mutation in NOD2 associated with susceptibility to Crohn's disease. Nature 2001;411(6837):603–6.
[13] Hugot JP, Chamaillard M, Zouali H, et al. Association of NOD2 leucine-rich repeat variants with susceptibility to Crohn's disease. Nature 2001;411(6837):599–603.
[14] Hisamatsu T, Suzuki M, Reinecker HC, et al. CARD15/NOD2 functions as an antibacterial factor in human intestinal epithelial cells. Gastroenterology 2003;124:993–1000.
[15] Wehkamp J, Harder J, Weichenthal M, et al. NOD2 (CARD15) mutations in Crohn's disease are associated with diminished mucosal alpha-defensin expression. Gut 2004; 53(11):1658–64.
[16] Kobayashi KS, Chamaillard M, Ogura Y, et al. Nod2-dependent regulation of innate and adaptive immunity in the intestinal tract. Science 2005;307(5710):731–4.
[17] Fedorak RN, Madsen KL. Probiotics and the management of inflammatory bowel disease. Inflamm Bowel Dis 2004;10(3):286–99.
[18] Shiba T, Aiba Y, Ishikawa H, et al. The suppressive effect of bifidobacteria on *Bacteroides vulgatus*, a putative pathogenic microbe in inflammatory bowel disease. Microbiol Immunol 2003;47(6):371–8.
[19] Mack DR, Ahrne S, Hyde L, et al. Extracellular MUC3 mucin secretion follows adherence of *Lactobacillus* strains to intestinal epithelial cells in vitro. Gut 2003;52(6):827–33.
[20] Freitas M, Tavan E, Cayuela C, et al. Host pathogens cross-talk. Indigenous bacteria and probiotics also play the game. Biol Cell 2003;95(8):503–6.
[21] Boudeau J, Glasser AL, Julien S, et al. Inhibitory effect of probiotic *Escherichia coli* strain Nissle 1917 on adhesion to and invasion of intestinal epithelial cells by adherent-invasive

E coli strains isolated from patients with Crohn's disease. Alimentary Pharmacol Ther 2003; 18(1):45–56.

[22] Madsen K, Cornish A, Soper P, et al. Probiotic bacteria enhance murine and human intestinal epithelial barrier function. Gastroenterology 2001;121(3):580–91.

[23] Resta-Lenert S, Barrett KE. Live probiotics protect intestinal epithelial cells from the effects of infection with enteroinvasive *Escherichia coli*. Gut 2003;52(7):988–97.

[24] Jijon H, Backer J, Diaz H, et al. DNA from probiotic bacteria modulates murine and human epithelial and immune function. Gastroenterology 2004;126(5):1358–73.

[25] Rachmilewitz D, Katakura K, Karmeli F, et al. Toll-like receptor 9 signaling mediates the anti-inflammatory effects of probiotics in murine experimental colitis. Gastroenterology 2004;126(2):520–8.

[26] Sheil B, McCarthy J, O'Mahony L, et al. Is the mucosal route of administration essential for probiotic function? Subcutaneous administration is associated with attenuation of murine colitis and arthritis. Gut 2004;53(5):694–700.

[27] Madsen KL, Doyle JS, Jewell LD, et al. *Lactobacillus* species prevents colitis in interleukin-10 gene-deficient mice. Gastroenterology 1999;116(5):1107–14.

[28] Schultz M, Veltkamp C, Dieleman LA, et al. *Lactobacillus plantarum* 299V in the treatment and prevention of spontaneous colitis in IL-10 deficient mice. Inflamm Bowel Dis 2002;8(2): 71–80.

[29] McCarthy J, O'Mahony L, O'Callaghan L, et al. Double blind, placebo controlled trial of two probiotic strains in interleukin 10 knockout mice and mechanistic link with cytokine balance. Gut 2003;52(7):975–80.

[30] Pena JA, Rogers AB, Ge Z, et al. Probiotic *Lactobacillus* spp. diminish *Helicobacter hepaticus*-induced inflammatory bowel disease in interleukin-10 deficient mice. Infect Immun 2005;73(2):912–20.

[31] Dieleman LA, Goerres MS, Arends A, et al. *Lactobacillus GG* prevents recurrence of colitis in HLA-B27 transgenic rats after antibiotic treatment. Gut 2003;52(3):370–6.

[32] Kennedy RJ, Hoper M, Deodhar K, et al. Probiotic therapy fails to improve gut permeability in a hapten model of colitis. Scand J Gastroenterol 2000;35(12):1266–71.

[33] Shibolet O, Karmeli F, Eliakim R, et al. Variable response to probiotics in two models of experimental colitis in rats. Inflamm Bowel Dis 2002;8(6):399–406.

[34] Schultz M, Strauch UG, Linde HJ, et al. Preventive effects of *Escherichia coli* strain Nissle 1917 on acute and chronic intestinal inflammation in two different murine models of colitis. Clin Diagn Lab Immunol 2004;11(2):372–8.

[35] Schotte L, Steidler L, Vandekerckhove J, et al. Secretion of biologically active murine interleukin-10 by *Lactococcus lactis*. Enzyme Microb Technol 2000;27(10):761–5.

[36] Steidler L, Hans W, Schotte L, et al. Treatment of murine colitis by *Lactococcus lactis* secreting interleukin-10. Science 2000;289(5483):1352–5.

[37] Ruseler-van Embden JG, Schouten WR, van Lieshout LM. Pouchitis: result of microbial imbalance? Gut 1994;35(5):658–64.

[38] Gionchetti P, Rizzello F, Venturi A, et al. Oral bacteriotherapy as maintenance treatment in patients with chronic pouchitis: a double-blind, placebo-controlled trial. Gastroenterology 2000;119(2):305–9.

[39] Mimura T, Rizzello F, Helwig U, et al. Once-daily high-dose probiotic therapy (VSL#3) for maintaining remission in recurrent or refractory pouchitis. Gut 2004;53(1):108–14.

[40] Gionchetti P, Rizzello F, Helwig U, et al. Prophylaxis of pouchitis onset with probiotic therapy: a double-blind, placebo-controlled trial. Gastroenterology 2003;124(5): 1202–9.

[41] Kuisma J, Mentula S, Jarvinen H, et al. Effect of *Lactobacillus rhamnosus* GG on ileal pouch inflammation and microbial flora. Dis Colon Rectum 2003;46(11):1476–83.

[42] Laake KO, Line PD, Aabakken L, et al. Assessment of mucosal inflammation and circulation in response to probiotics in patients operated with ileal pouch anal anastomosis for ulcerative colitis. Scand J Gastroenterol 2003;38(4):409–14.

[43] Schultz M, Timmer A, Herfarth HH. *Lactobacillus GG* in inducing and maintaining remission of Crohn's disease. BMC Gastroenterol 2004;4(1):5.
[44] McCarthy J, O'Mahony L, Dunne C, et al. An open trial of a novel probiotic as an alternative to steroids in mild/moderately active Crohn's disease. Gut 2001;49(Suppl 3):A2447.
[45] Gupta P, Andrew H, Kirschner BS, et al. Is *Lactobacillus GG* helpful in children with Crohn's disease? Results of a preliminary, open-label study. J Ped Gastroenterol Nutr 2000; 31(4):453–7.
[46] Malchow HA. Crohn's disease and *Escherichia coli*. A new approach in therapy to maintain remission of colonic Crohn's disease? J Clin Gastroenterol 1997;25(4):653–8.
[47] Guslandi M, Mezzi G, Sorghi M, et al. *Saccharomyces boulardii* in maintenance treatment of Crohn's disease. Dig Dis Sci 2000;45:1462–4.
[48] Prantera C, Scribano ML, Falasco G, et al. Ineffectiveness of probiotics in preventing recurrence after curative resection for Crohn's disease: a randomised controlled trial with *Lactobacillus GG*. Gut 2002;51(3):405–9.
[49] Campieri M, Rizzello F, Venturi A, et al. Combination of antibiotic and probiotic treatment is efficacious in prophylaxis of postoperative recurrence of Crohn's disease: a randomized controlled study versus mesalamine. Gastroenterology 2000;118:A781.
[50] Rutgeerts P, Hiele M, Geboes K, et al. Controlled trial of metronidazole treatment for prevention of Crohn's recurrence after ileal resection. Gastroenterology 1995;108(6): 1617–21.
[51] Kato K, Mizuno S, Umesaki Y, et al. Randomized placebo-controlled trial assessing the effect of *bifidobacteria*-fermented milk on active ulcerative colitis. Aliment Pharmacol Ther 2004;20(10):1133–41.
[52] Rembacken BJ, Snelling AM, Hawkey PM, et al. Nonpathogenic *Escherichia coli* versus mesalazine for the treatment of ulcerative colitis: a randomised trial. Lancet 1999; 354(9179):635–9.
[53] Fedorak RN, Gionchetti P, Campieri M, et al. VSL3 probiotic mixture induces remission in patients with active ulcerative colitis. Gastroenterol 2003;124:A377.
[54] Guslandi M, Giollo P, Testoni PA. A pilot trial of *Saccharomyces boulardii* in ulcerative colitis. Eur J Gastroenterol Hepatol 2003;15(6):697–8.
[55] Borody TJ, Warren EF, Leis SM. Treatment of ulcerative colitis using fecal bacteriotherapy. J Clin Gastroenterol 2003;37(1):42–7.
[56] Kruis W, Fric P, Pokrotnieks J, et al. Maintaining remission of ulcerative colitis with the probiotic *Escherichia coli* Nissle 1917 is as effective as with standard mesalazine. Gut 2004;53(11):1617–23.
[57] Ishikawa H, Akedo I, Umesaki Y, et al. Randomized controlled trial of the effect of bifidobacteria-fermented milk on ulcerative colitis. J Am Coll Nutr 2003;22(1):56–63.
[58] Venturi A, Gionchetti P, Rizzello F, et al. Impact on the composition of the faecal flora by a new probiotic preparation: preliminary data on maintenance treatment of patients with ulcerative colitis. Aliment Pharmacol Ther 1999;13(8):1103–8.
[59] Gibson GR, Roberfroid MB. Dietary modulation of the human colonic microbiota: introducing the concept of prebiotics. J Nutr 1995;125:1401–12.
[60] Langlands SJ, Hopkins MJ, Coleman N, et al. Prebiotic carbohydrates modify the mucosa associated microflora of the human large bowel. Gut 2004;53(11):1610–6.
[61] Araki Y, Andoh A, Koyama S, et al. Effects of germinated barley foodstuff on microflora and short-chain fatty acid production in dextran sulfate sodium-induced colitis in rats. Biosci Biotechnol Biochem 2000;64(9):1794–800.
[62] Moreau NM, Martin LJ, Toquet CS, et al. Restoration of the integrity of the rat caeco–colonic mucosa by resistant starch but not by fructo–oligosaccharides, in dextran sulfate sodium-induced experimental colitis. Br J Nutr 2003;90:75–85.
[63] Holma R, Juvonen P, Asmawi MZ, et al. Galacto–oligosaccharides stimulate the growth of bifidobacteria but fail to attenuate inflammation in experimental colitis in rats. Scand J Gastroenterol 2002;37:1042–7.

[64] Kanauchi O, Mitsuyama K, Homma T, et al. Treatment of ulcerative colitis patients by long-term administration of germinated barley foodstuff: multi-center open trial. Int J Mol Med 2003;12(5):701–4.
[65] Hanai H, Kanauchi O, Mitsuyama K, et al. Germinated barley foodstuff prolongs remission in patients with ulcerative colitis. Int J Mol Med 2004;13(5):643–7.
[66] Furrie E, MacFarlane S, Kennedy A, et al. Synbiotic therapy (*Bifidobacterium longum*/ Synergy 1) initiates resolution of inflammation in patients with active ulcerative colitis: a randomized controlled pilot trial. Gut 2005;54:242–9.
[67] Pool-Zobel B, van Loo J, Rowland I, et al. Experimental evidences on the potential of prebiotic fructans to reduce the risk of colon cancer. Br J Nutr 2002;87(Suppl 2):S273–81.

Gastroenterol Clin N Am 34 (2005) 483–498

ELSEVIER SAUNDERS

GASTROENTEROLOGY CLINICS
OF NORTH AMERICA

Lactobacillus GG: Bacteriology and Clinical Applications

Shira Doron, MD[a,*], David R. Snydman, MD[a], Sherwood L. Gorbach, MD[b,†]

[a]Department of Medicine, Tufts–New England Medical Center, 750 Washington Street, NEMC #041, Boston, MA 02111, USA
[b]Department of Medicine, Tufts University School of Medicine, 150 Harrison Avenue, Jaharis 2, Boston, MA 02111, USA

Attempts over many years to implant various dairy strains of *Lactobacillus* into the human intestine have proved unsuccessful. Researchers concluded that the *Lactobacillus* strains traditionally used in the dairy industry, which included *L casei, L bulgaricus,* and *L acidophilus,* did not possess the biologic characteristics necessary to enable them to implant consistently in the human intestine and would be unlikely to have a beneficial effect on human health. Based on this analysis, Gorbach and Goldin started by listing the critical properties of an ideal *Lactobacillus* strain that would benefit human health, then set out to isolate it [1]. They collected strains of *Lactobacillus* from stool specimens obtained from healthy human volunteers and submitted the candidate strains to a series of tests. The strain needed to be stable in stomach acid and bile to survive the upper gastrointestinal tract on its way to the colon. The strains that were most resistant to acid and bile were tested for adherence to human intestinal epithelial cells. Finally, the strains that adhered well to intestinal cells were tested for production of an antimicrobial substance active against various pathogenic and commensal bacteria. The strains producing the largest amount of antibacterial substance were assayed to assess their growth rate. The strain with the fastest growth rate that possessed these other attributes was identified in 1985 and was named *Lactobacillus rhamnosus* GG (LGG) for its discoverers. They also noted that LGG has a distinct colonial morphology that makes it easy to identify in a mixed culture of other lactobacilli, such as stool and oropharyngeal flora. The criteria used to isolate this strain (Box 1) subsequently have been used to develop members of the newer generation of

†Dr. Gorbach is one of the discoverers of LGG and holds a patent on the organism, which is licensed to Valio, Ltd. He also derives royalties from sales of LGG products.

*Corresponding author. *E-mail address*: sdoron@tufts-nemc.org (S. Doron).

0889-8553/05/$ – see front matter
doi:10.1016/j.gtc.2005.05.011

gastro.theclinics.com

Box 1: Criteria for an ideal probiotic strain

Resistance to acid and bile
Attachment to human epithelial cells
Colonize the human intestine
Produce an antimicrobial substance
Good growth characteristics
Beneficial effects of human health

probiotics. A patent was issued in the United States in 1987 and subsequently in other countries. Valio, Ltd., in Finland was granted the initial license, and since then LGG dairy products, juices, cheeses, and capsules have been marketed in more than 40 countries worldwide. LGG is popular in Finland, with an estimated annual per capita consumption rate that has increased from approximately 1 L in 1995 to 6 L in 2000 [2]. In 1992, it was estimated that 3 million kg of LGG were consumed by a minimum of 40,000 persons in Finland alone [3].

BACTERIOLOGY

LGG is a gram-positive rod that forms a creamy white colony on a culture plate and has a distinct buttery smell. It can be distinguished on Gram stain from other lactobacilli by its palisading appearance. It ferments xylose, trehalose, sorbitol, salicilin, ribose, rhamnose, melezitose, mannose, mannitol, glucose, fructose, and cellobiose. It does not ferment amygdalin, arabinose, erythritol, glycogen, inositol, lactose, maltose, melibiose, raffinose, or sucrose. The electrophoretic pattern of LGG-soluble proteins is similar to that of *L casei* subsp *rhamnosus,* but LGG differs from that strain in that it cannot ferment lactose, maltose, or sucrose. LGG produces an inhibitory substance, probably a short-chain fatty acid, with activity against a variety of bacterial species, including anaerobic bacteria (*Clostridium, Bacteroides,* and *Bifidobacterium*), *Escherichia coli, Pseudomonas, Staphylococcus, Streptococcus,* and *Salmonella* [4].

LGG has powerful adhesive properties. In vitro using differentiated human colon adenocarcinoma Caco-2 cells, LGG has one of the highest binding capacities among the lactobacilli [5,6].

After ingestion by humans, LGG can be cultured in stool for 7 days [7] and from intestinal mucosa biopsy specimens for 28 days [8]. LGG also can be cultured from saliva for 2 weeks after ingestion [9]. In healthy volunteers given LGG, changes in stool microecology consist of increased concentrations of bifidobacteria and lactobacilli and decreased concentrations of lecithinase-negative clostridia [10]. Compared with milk-hypersensitive adults (a group in which clinical benefits have been shown from treatment with LGG [11]), normal subjects given LGG have a greater increase in fecal bifidobacteria [12]. In neonates treated with LGG, colonization is more likely to occur in an infant with a birth weight greater than 1500 g [13]. In a study of infants of women who took LGG during pregnancy, most infants themselves were colonized with LGG, some for 24 months after birth [14].

RISK OF BACTEREMIA WITH *LACTOBACILLUS* GG

Because of the high rate of consumption of LGG in Finland, researchers examined all *Lactobacillus* blood isolates between 1990 and 2000. Eleven of 89 *Lactobacillus* bacteremia isolates could not be distinguished from LGG by pulsed-field gel electrophoresis; however, despite increasing consumption of LGG, the rate of *Lactobacillus* bacteremia remained constant over time [15]. The researchers concluded that LGG bacteremia was not a risk in Finnish people who consumed large amounts of LGG dairy products. In a second study, the same researchers examined the LGG-like clinical isolates for nongenetic phenotypic properties that might distinguish them from the probiotic strain. When comparing probiotic LGG with the clinical *Lactobacillus* isolates, they found differences in their in vitro adhesion properties, so it is unlikely that the bacteremic strains were LGG [2].

HEALTH EFFECTS

Colonization by *Lactobacillus* GG during diarrhea

Competitive colonization is one of the main reasons for the effectiveness of LGG in intestinal infections. LGG binds to enteric epithelium, inhibiting adhesion of pathogens; this has been shown in vitro with *E coli* O157:H7 [16,17]. LGG also may compete with other bacteria for consumption of monosaccharides, slowing their growth. This characteristic is likely to be partially responsible for the effectiveness of LGG in *Clostridium difficile* colitis [18]. Other probiotics have been shown to modify toxin receptors, specifically *C difficile* toxin A, through an enzymatic process [19].

Fecal concentrations of short-chain fatty acids

Lactobacilli produce short-chain fatty acids, which lower the colonic pH, favoring the growth of less pathogenic organisms.

Modulation of antigen transport

The role of LGG as a modulator of antigen transport may explain its effects in preventing food allergies in children. LGG appeared to close molecular pores in the intestines of rats given LGG compared with milk or water. Because intestinal inflammation may allow antigens to cross the mucosal barrier, leading to sensitization of the immune system, the closing of permeability gaps may protect against the development of food allergies in susceptible infants [20]. LGG given to germ-free rats increases the rate of mitoses in the intestinal epithelial cells [21].

Augmentation of the immune response

Nonspecific augmentation of the immune response is likely to be the reason why LGG is effective against such a wide variety of infections. In a study of children with acute rotavirus diarrhea, researchers were able to show an antigen-nonspecific immunoglobulin response, with IgM being especially prominent, that was almost twice as pronounced in the LGG group as that in the placebo group. This response likely explains the shortening of the duration of gastroenteritis symptoms seen in rotavirus patients given LGG. During the convalescent phase, the patients who had been given LGG had significantly higher levels of rotavirus-specific IgA; this may contribute to protection against reinfection with rotavirus [22].

LGG has been shown to enhance various immune responses. In one study in healthy volunteers [23], LGG heightened the activational response of peripheral $CD4^+$ T lymphocytes to intestinal bacterial components and enhanced an anti-inflammatory type of response by causing increased secretion of suppressive cytokines (interleukin-10, interleukin-4) and decreased secretion of proinflammatory cytokines (tumor necrosis factor-α, interleukin-6, interferon-γ). In another study, milk increased expression of phagocytosis receptors in milk-hypersensitive adults, whereas normal volunteers showed no such response. When these same groups of subjects were given LGG, expression of phagocytosis receptors was stimulated in the normal subjects and attenuated in the milk-hypersensitive group. The results of this study suggest that LGG can have bidirectional immunomodulatory effects–stimulatory when an increased response is needed and down-regulatory when attenuation is appropriate [11].

BENEFITS OF *LACTOBACILLUS* GG

The benefits of LGG can be classified as proven benefits (in two or more comparative trials), benefits for which there is substantial evidence but additional proof is needed, benefits related to ongoing research in promising areas, and benefits related to future areas of research (Box 2).

Proven benefits

Treatment and prevention of acute diarrhea in children

Many studies suggest that LGG is helpful in the treatment of acute diarrhea in children [24]. Childhood diarrhea is common throughout the world, but in the developing world it is twice as common and is associated with weight loss and significant mortality. The largest study of LGG for acute diarrhea in children was conducted by the European Society for Pediatric Gastroenterology, Hepatology and Nutrition. In this study, 287 children in 10 countries age 1 to 36 months admitted with moderate-to-severe diarrhea were randomized to receive either LGG or placebo along with standard oral rehydration solution. The patients who received LGG had a shorter duration and decreased severity of illness, had a shorter hospital stay, and were less likely to have persistent diarrhea [25]. In another study in the Karelian Republic, 137 children age 1 to 36 months admitted with diarrhea were treated with LGG or placebo plus oral rehydration solution. LGG significantly shortened the duration of illness [26]. In Thailand, Pant et al [27] studied 26 children with watery diarrhea. The mean duration of diarrhea was significantly shorter in the group treated with

Box 2: Benefits of LGG

Proven benefits
Treatment and prevention of acute diarrhea in children

Prevention of antibiotic-associated diarrhea

Substantial evidence but needs additional proof
Prevention and treatment of allergies

Prevention of traveler's diarrhea

Prevention of respiratory infections in children in daycare

Prevention of dental caries

Elimination of nasal colonization with potentially pathogenic bacteria

Ongoing research in promising areas
Treatment of inflammatory bowel disease and prevention of relapses

Treatment of intestinal inflammation in cystic fibrosis

Treatment of *Clostridium difficile*

Future areas of research
Human studies

- Treatment of rheumatoid arthritis
- Treatment of irritable bowel syndrome

Animal and in vitro studies

- Prevention of cancer
- Prevention of complications of liver disease
- Prevention of complications of radiation exposure
- Treatment of diabetes
- Treatment of graft-versus-host disease

LGG. In Pakistan, Raza et al [28] found LGG to be effective as well. Children with diarrhea ($n = 40$) received LGG or placebo. Children treated with LGG were less likely to have persistent diarrhea and had fewer episodes of vomiting per day. An Estonian study [29] looked at children with diarrhea caused by *Shigella sonnei*. Children ($n = 31$) received trimethoprim-sulfamethoxazole with or without LGG. The patients in the LGG group had a significantly higher cure rate by day 10. Promising results showing usefulness of LGG in acute diarrhea in children, which is caused mostly by rotavirus, led some researchers to examine its effectiveness in preventing the spread of rotavirus diarrhea within hospitals. In one study of 81 children age 1 to 36 months hospitalized for reasons other than diarrhea, symptoms of nosocomial rotavirus gastroenteritis were prevented by administration of LGG, whereas actual rates of detection of rotavirus in stool were comparable for LGG-treated and untreated subjects [30]. In another much larger study [31] of 220 patients, there was no difference in rates of nosocomial rotavirus in the LGG-treated versus placebo group.

LGG has been shown not only to treat, but also to prevent diarrhea in children in developing countries. In a Peruvian study, 204 undernourished children age 6 to 24 months were randomized to LGG versus placebo. There was a significant reduction in the incidence of diarrhea in the LGG group, largely limited to non-breastfed children [32]. Results of studies are convincing for the utility of LGG in the treatment of acute diarrhea in children, but its use in prevention of diarrhea requires further investigation.

Prevention of antibiotic-associated diarrhea

Several studies have shown that LGG can reduce the intestinal side effects and diarrhea associated with the use of antibiotics [33–37]. Arvola et al [33] randomized 119 children taking antibiotics for respiratory infections to LGG versus placebo and found a reduction in diarrhea in the LGG group to one third its incidence in the placebo group during the first 2 weeks after beginning antibiotics. Vanderhoof et al [34] randomized 202 children on oral antibiotics to LGG versus placebo and found a reduction in incidence of diarrhea from 26% to 8%. Armuzzi et al [35,37] performed two randomized studies, one with 60 adults and one with 120 adults receiving antibiotic therapy for *Helicobacter pylori* plus either LGG or placebo. Significantly less nausea, taste disturbance, and diarrhea were noted in the LGG groups. A Finnish group gave 16 healthy volunteers a generous dose of erythromycin, which is notorious for its gastrointestinal side effects, plus LGG-containing yogurt or pasteurized yogurt as placebo. They showed that LGG prevented a variety of gastrointestinal symptoms caused by erythromycin, including abdominal pain, diarrhea, and nausea [38].

Substantial evidence but needs additional proof

Prevention and treatment of allergies

Atopic eczema, allergic rhinitis, and asthma all are manifestations of allergy that seem to be increasing in prevalence in the developed world. Some researchers blame improved hygiene on the increase. Increasingly, scientists are looking at the role of commensal intestinal flora as modulators of immune function and regulators of the response to antigens. Kalliomaki et al [39] randomized 159 pregnant women with a family history of atopic disease to LGG capsules or placebo for 2 to 4 weeks before expected delivery. Women who chose to breastfeed continued to take the LGG or placebo for 6 months, whereas women who did not breastfeed gave the LGG or placebo directly to their infants. The frequency of atopic eczema was reduced by half over the course of the first 2 years of life in the LGG group. Breast milk was assayed from the 62 mothers who breastfed, and the mothers in the LGG group had significantly higher levels of transforming growth factor-β_2 in the milk [40]. Four years after birth, this cohort was assessed for persistence of the preventive effect of LGG on atopic disease. The same findings persisted with a significantly larger percentage of patients in the placebo group developing atopic eczema compared with the LGG group [41].

LGG has been shown to be effective not only for prevention, but also for treatment of allergy manifestations. In one study, 27 infants with atopic eczema were randomized to receive LGG, another probiotic, or placebo. The probiotic-treated groups had a significant improvement in skin condition after 2 months compared with the placebo group [42]. In another treatment study, 31 infants with atopic eczema and cow's milk allergy were put on a cow's milk elimination diet plus LGG or placebo for 1 month. There was a significant improvement in eczema in the patients receiving LGG, but not in patients receiving placebo. In a second part of the same study, mothers of 11 breastfed infants with atopic eczema took LGG. These infants also had a significant improvement in eczema at the end of the treatment period [43]. The evidence is convincing that LGG can mediate the manifestations of atopy in infants.

Prevention of traveler's diarrhea

Travelers to underdeveloped countries are subject to a high incidence of diarrhea, ranging from 25% to 50%. A large Finnish study successfully randomized 756 subjects traveling to two resorts in Turkey to take LGG or placebo during their stay abroad. The overall rate of diarrhea was 43.8%. The results from subjects at one of the resorts were not statistically different between groups. At the other resort, the LGG group had a statistically significant protection rate of 39.5% in the first week and 27.9% in the second week [44]. In another study, 245 travelers from New York who journeyed to developing countries for periods of 1 to 3 weeks were given LGG versus placebo to take during their travels. Subjects kept daily diaries documenting health symptoms, including intestinal complaints. LGG reduced the risk of diarrhea on any given day from 7.4% to 3.9%, a protection rate of 47% [45]. Further studies are needed to evaluate whether LGG might be a useful preventive measure for travelers at risk for diarrhea.

Prevention of respiratory infections in children in daycare

Children in daycare have a high rate of respiratory and gastrointestinal infections. Researchers examined the effects of long-term consumption of probiotic milk over the course of a winter in Helsinki. Children were randomized to LGG or placebo to be consumed at the daycare centers three times a day, 5 days a week, for 7 months. There were 571 children age 1 to 6 in the study. Children in the LGG group had modest reductions in days of absence owing to illness from daycare, respiratory infections with complications, lower respiratory tract infections, and antibiotic treatments for respiratory infection [46]. Further studies are needed to define possible benefits in preventing respiratory illnesses.

Prevention of dental caries

Given the ability of LGG to colonize the mouth and its presence in the saliva of individuals who ingest it, researchers conducted a study of the ability of LGG to prevent dental caries in children. LGG has activity against *Streptococcus* species, thought to be major players in the development of caries. The daycare children in the aforementioned large multicenter trial on prevention of respiratory infection in daycare were examined by dentists before and after the 7-month intervention. The LGG milk had a significantly beneficial effect on clinical development of dental caries, particularly in the subgroup of children age 3 to 4 years [47]. In addition, the viable counts of *Streptococcus mutans,* the putative cause of dental caries, were significantly reduced in the mouths of children in the LGG group.

Elimination of nasal colonization with potentially pathogenic bacteria

In a study by Gluck and Gebbers [48], 209 healthy volunteers were randomized to receive either a fermented milk product with probiotics including LGG or a yogurt drink placebo. Subjects' nares were swabbed for potentially pathogenic bacteria. A significantly larger proportion of subjects receiving LGG eliminated gram-positive bacteria, including *Staphylococcus aureus, Streptococcus pneumoniae,* and beta-hemolytic streptococci, compared with placebo-treated individuals.

Ongoing research in promising areas

Treatment of inflammatory bowel disease and prevention of relapses

Scientists only more recently have begun to understand the importance of the interaction between commensal gut flora and the normal immune response within the intestine. Inflammatory bowel disease (IBD), including ulcerative colitis and Crohn's disease, is characterized by dysregulation of the immune system leading to inflammation of the gastrointestinal tract. Patients with ulcerative colitis have been shown to have an excess of enteroadhesive and enterohemorrhagic *E coli* [49], and Crohn's disease patients have been found to have a decreased concentration of bifidobacteria in their stool [50]. For these reasons, and given the evidence that LGG leads to closing of intestinal permeability gaps as discussed earlier, the idea of giving an oral bacterial supplement to IBD patients has been an attractive one.

Data on the effectiveness of LGG in IBD are conflicting. A pilot study of four patients with Crohn's disease treated with LGG for 6 months suggested clinical improvement accompanied by decreased intestinal permeability during treatment with LGG [51]. In a study of ulcerative colitis patients enrolled just after ileal pouch–anal anastomosis, first episode of pouchitis (inflammation of the ileal pouch–anal anastomosis, which occurs in 20–59% of patients [52]) was delayed by administration of LGG [53]. In a similar study by the same authors [52], 20 patients with chronic pouchitis were randomized to LGG or placebo. Only one patient, who became heavily colonized with LGG, showed clinical improvement. In an open study of LGG in children with Crohn's disease, no clinical improvement was seen despite stimulation of secretion of IgA, the gut's immunologic barrier [54]. A study looking at postoperative recurrence of Crohn's disease similarly failed to show an effect of LGG over the course of 1 year [55].

Treatment of intestinal inflammation in cystic fibrosis
Chronic abdominal pain is a characteristic of cystic fibrosis, and there is evidence that it is due to intestinal inflammation. Researchers in Italy documented intestinal inflammation in cystic fibrosis patients using the markers fecal calprotectin and rectal nitric oxide. Both of these markers were reduced by administration of LGG [56].

Treatment of Clostridium difficile
C difficile diarrhea, caused in most cases by a disturbance of intestinal flora secondary to antibiotic use, has a high recurrence rate after treatment with metronidazole or vancomycin, the drugs of choice. Research has shown that *C difficile* can respond to microbial replacement therapy, such as with nontoxigenic *C difficile* and fecal enemas; however, these treatments are difficult to carry out and undesirable to most patients. After cohorts of 4 children [57] and 5 adults [58] with relapsing diarrhea caused by *C difficile* were treated successfully with LGG, researchers from Tufts and Johns Hopkins treated 32 patients with relapsing *C difficile* with LGG for 21 days. Although more than 60% of patients with relapsing *C difficile* would be expected, based on prior experience, to experience another episode of *C difficile* within 4 weeks after a course of standard therapy with metronidazole or vancomycin, in this cohort only 16% (5 of 32) had another relapse after a single course of LGG, and an additional 3 patients were cured with subsequent courses of LGG for an overall cure rate of 94% [59]. Proper controlled trials of LGG in this disease have not been performed.

Future areas of research

Human studies

Treatment of rheumatoid arthiritis. Similar to IBD, rheumatoid arthritis is a chronic systemic inflammatory disease. Some studies have suggested that rheumatoid arthritis patients have abnormal fecal flora, and that changes in the fecal flora can be tied to arthritis activity [60,61]. Experimental arthritis in Lewis rats was significantly improved with LGG compared with placebo [62]. Hatakka et al [63] conducted a randomized double-blind, placebo-controlled trial of LGG versus placebo in 21 rheumatoid arthritis patients. Patients in the LGG group had a decrease in the number of swollen joints and overall rheumatoid arthritis activity, but these differences were not statistically significant.

Treatment of irritable bowel syndrome. Irritable bowel syndrome (IBS) is a common disease affecting approximately 12% of United States adults [64]. Characterized by abdominal pain and altered bowel function, its cause is not completely understood. Theories include altered gut reactivity, enhanced perception of intestinal distention and pain, and dysregulation of the brain-gut axis. Inflammation and immune dysfunction also are thought to play a role [65,66]. There is strong evidence for a relationship between commensal bacteria and intestinal health, with theories suggesting IBS patients have altered gastrointestinal flora [67]. After several studies showed effectiveness of other probiotics in alleviating IBS symptoms, O'Sullivan and O'Morain [68] randomized 24 IBS patients in a 20-week, double-blind crossover study of LGG versus placebo. LGG did not significantly improve the symptoms of bloating, pain, or urgency in this small cohort of IBS patients. A downward trend in pain score was noted, however, on LGG treatment. In the diarrhea group, when considered separately, there was also a downward trend in the number of unformed bowel movements on LGG treatment. The authors concluded that further investigation of the efficacy of LGG in IBS is warranted [68].

Future areas of research

Animal and in vitro studies

Prevention of cancer. A model of colon cancer in rats was induced by dimethylhydrazine to determine whether LGG could slow the development of intestinal tumors. The rats given LGG before dimethylhydrazine had a significantly lower incidence of colon tumors compared with rats not supplemented with LGG. In LGG-treated animals that did develop tumors, the number of tumors was smaller than that among control animals [69].

In mice inoculated with bladder cancer cells, consumption of LGG prevented or significantly slowed development of tumors. The LGG was most efficacious when administered immediately after inoculation of cancer cells, when the tumor burden was smallest. The researchers postulated that the antitumor effects of LGG may be due to general immune activation, stimulation of cytotoxic immune cells against bladder tumor cells, or transport of LGG or its products to the tumor site with a direct local killing effect or induction of local cytokine production [70].

In vitro studies show that LGG can bind aflatoxin, a potent food-borne carcinogen [71,72]. This too may have implications as to the ability of long-term consumption of LGG to prevent cancer. Although these various studies are promising, it remains for properly conducted clinical trials to establish whether LGG has any role in prevention of cancer.

Prevention of complications of liver disease. Knowing that elimination of intestinal bacteria by colectomy in rats [73] and enterectomy in rabbits [74] prevents chemically induced liver injury, and given the effectiveness of gut-sterilizing agents, such as neomycin and polymyxin B, in humans [75], Nanji et al [76] sought to alter intestinal flora in a rat model of liver disease by administration of LGG. The researchers induced liver injury in rats using ethanol and supplemented one group with LGG. Rats fed LGG had a reduction in liver injury and a lower plasma level of endotoxin. The results of this study hold promise for patients with alcoholic liver disease. Bauer et al [77] used rats with carbon tetrachloride–induced liver disease as a model of chronic liver disease to study the effect of LGG on the prevention of spontaneous bacterial peritonitis, a problem that plagues humans with this condition. The investigators were unable to show a decrease in intestinal bacterial overgrowth, bacterial translocation into mesenteric lymph nodes or portal blood, or positive ascitic fluid cultures, all of which occur with substantial frequency in this rat model of cirrhosis [77].

Prevention of complications of radiation exposure. Mice colonized with LGG and controls were subjected to high-dose total-body irradiation (1400 rad). The LGG-fed mice had a significantly decreased mortality rate at 48 hours. All of the mice that died had bacteremia. Although gram-negative organisms predominated in the control group, the LGG-fed mice were more likely to have had streptococcal bacteremia. Ten of 21 controls had *P aeruginosa* bacteremia in contrast to 1 of 21 LGG-fed mice. None of the mice had *Lactobacillus* isolated from blood cultures [78]. The results of this study have implications for humans with numerous conditions causing injury to the bowel, but also underscore the safety of LGG when administered even to individuals with disrupted intestinal mucosa.

Treatment of diabetes. Using a rat model of diabetes, a group of Japanese researchers showed that LGG has the capability to lower blood hemoglobin A_{1c} and improve glucose tolerance test results compared with controls. The authors of the study suggest that the mechanism may be use of glucose by the lactobacilli [79]. Further studies are needed.

Treatment of graft-versus-host disease. Bacterial lipopolysaccharide is thought to play a major role in the development of acute graft-versus-host disease (GVHD) in bone marrow transplantation patients [80], and bowel decontamination using broad-spectrum antibiotics has been used before transplant to avoid this complication [81]. Gerbitz et al [82] studied a mouse model of acute GVHD in which the disease is induced by major histocompatibility mismatch. The animals generally have severe damage to bowel mucosa and high levels of serum lipopolysaccharide and proinflammatory cytokines. In this study, mice were given drinking water containing LGG, ciprofloxacin, or no additive beginning 7 days before transplantation. Treatment with LGG significantly reduced mortality, particularly in the early post-transplantation period and led to a reduced GVHD score compared with the other groups. The improvement in GVHD score was mostly due to a reduction in weight loss and to better skin integrity. Histopathologic bowel inflammation was found to be less in the LGG-treated mice. T cell proliferation in the spleen was reduced in mice treated with LGG, as evidenced by lower numbers of $CD3^+$ cells. Splenocytes of LGG-treated mice stimulated with lipopolysaccharide had a lower spontaneous release of interferon-γ. Mesenteric lymph nodes of LGG-treated animals had a lower concentration of translocated gut organisms, but this did not translate into lower serum lipopolysaccharide levels in these animals [82].

SUMMARY

LGG has been studied extensively in humans and experimental animals for a wide variety of uses. In some cases, data are conflicting and limited by small sample sizes. Nevertheless, the many possible indications for its use are bound to spawn continued enthusiasm for LGG clinical trials. At present, there is strong evidence that LGG can be used in acute diarrhea and antibiotic-associated diarrhea. More data are needed on the use of LGG in atopic disease and for prevention of infections in daycare, traveler's diarrhea, and dental caries. Results from studies in IBD, *C difficile,* and cystic fibrosis are promising. Finally, studies of LGG for cancer prevention, rheumatoid arthritis, IBS, liver disease, and diabetes warrant more research. Altering the intestinal flora with probiotics is an exciting approach to managing intestinal disorders and related conditions. LGG and other products are safe, cheap, and easy to administer. Future investigators will be challenged to define their utility in treatment and prevention of the broad array of potential clinical settings.

References

[1] Gorbach SL. The discovery of *Lactobacillus* GG. Nutr Today 1996;31:25–45.

[2] Ouwehand AC, Saxelin M, Salminen S. Phenotypic differences between commercial *Lactobacillus rhamnosus* GG and *L. rhamnosus* strains recovered from blood. Clin Infect Dis 2004;39:1858–60.

[3] Saxelin M. LGG Summatim: *Lactobacillus* GG and its health effects. 2nd edition. Helsinki (Finland): Valio Ltd, R&D; 2002.

[4] Silva M, Jacobus NV, Deneke C, Gorbach SL. Antimicrobial substance from a human *Lactobacillus* strain. Antimicrob Agents Chemother 1987;31:1231–3.

[5] Chauviere G, Coconnier MH, Kerneis S, Fourniat J, Servin AL. Adhesion of human *Lactobacillus acidophilus* strain LB to human enterocyte-like Caco-2 cells. J Gen Microbiol 1992;138(Pt 8):1689–96.

[6] Lehto EM, Salminen SJ. Inhibition of *Salmonella typhimurium* adhesion to Caco-2 cell cultures by *Lactobacillus* strain GG spent culture supernate: only a pH effect? FEMS Immunol Med Microbiol 1997;18:125–32.

[7] Goldin BR, Gorbach SL, Saxelin M, Barakat S, Gualtieri L, Salminen S. Survival of *Lactobacillus* species (strain GG) in human gastrointestinal tract. Dig Dis Sci 1992;37: 121–8.

[8] Alander M, Satokari R, Korpela R, et al. Persistence of colonization of human colonic mucosa by a probiotic strain, *Lactobacillus rhamnosus* GG, after oral consumption. Appl Environ Microbiol 1999;65:351–4.

[9] Meurman JH, Antila H, Salminen S. Recovery of *Lactobacillus* strain GG (ATCC 53103) from saliva of healthy volunteers after consumption of yoghurt prepared with the bacterium. Microb Ecol Health Dis 1994;7:295–8.

[10] Benno Y, He F, Hosoda M, et al. Effects of *Lactobacillus* GG yogurt on human intestinal microecology in Japanese subjects. Nutr Today Suppl 1996;31:9S–11S.

[11] Pelto L, Isolauri E, Lilius EM, Nuutila J, Salminen S. Probiotic bacteria down-regulate the milk-induced inflammatory response in milk-hypersensitive subjects but have an immunostimulatory effect in healthy subjects. Clin Exp Allergy 1998;28:1474–9.

[12] Apostolou E, Pelto L, Kirjavainen PV, Isolauri E, Salminen SJ, Gibson GR. Differences in the gut bacterial flora of healthy and milk-hypersensitive adults, as measured by fluorescence in situ hybridization. FEMS Immunol Med Microbiol 2001;30:217–21.

[13] Agarwal R, Sharma N, Chaudhry R, et al. Effects of oral *Lactobacillus* GG on enteric microflora in low-birth-weight neonates. J Pediatr Gastroenterol Nutr 2003;36:397–402.

[14] Schultz M, Gottl C, Young RJ, Iwen P, Vanderhoof JA. Administration of oral probiotic bacteria to pregnant women causes temporary infantile colonization. J Pediatr Gastroenterol Nutr 2004;38:293–7.

[15] Salminen MK, Rautelin H, Tynkkynen S, et al. *Lactobacillus* bacteremia, clinical significance, and patient outcome, with special focus on probiotic *L. rhamnosus* GG. Clin Infect Dis 2004;38:62–9.
[16] Mack DR, Michail S, Wei S, McDougall L, Hollingsworth MA. Probiotics inhibit enteropathogenic *E. coli* adherence in vitro by inducing intestinal mucin gene expression. Am J Physiol 1999;276(4 Pt 1):t-50.
[17] Michail S, Wei S, Mack DR. *Escherichia coli* strain E2348/69 in vitro adhesion is reduced in the presence of a *Lactobacillus* species. Gastroenterology 1997;112:A1042.
[18] Wilson KH, Perini F. Role of competition for nutrients in suppression of *Clostridium difficile* by the colonic microflora. Infect Immun 1988;56:2610–4.
[19] Pothoulakis C, Kelly CP, Joshi MA, et al. *Saccharomyces boulardii* inhibits *Clostridium difficile* toxin A binding and enterotoxicity in rat ileum. Gastroenterology 1993;104: 1108–15.
[20] Isolauri E, Majamaa H, Arvola T, Rantala I, Virtanen E, Arvilommi H. *Lactobacillus casei* strain GG reverses increased intestinal permeability induced by cow milk in suckling rats. Gastroenterology 1993;105:1643–50.
[21] Banasaz M, Norin E, Holma R, Midtvedt T. Increased enterocyte production in gnotobiotic rats mono-associated with *Lactobacillus rhamnosus* GG. Appl Environ Microbiol 2002; 68:3031–4.
[22] Kaila M, Isolauri E, Soppi E, Virtanen E, Laine S, Arvilommi H. Enhancement of the circulating antibody secreting cell response in human diarrhea by a human *Lactobacillus* strain. Pediatr Res 1992;32:141–4.
[23] Schultz M, Linde HJ, Lehn N, et al. Immunomodulatory consequences of oral administration of *Lactobacillus rhamnosus* strain GG in healthy volunteers. J Dairy Res 2003;70:165–73.
[24] Allen SJ, Okoko B, Martinez E, Gregorio G, Dans LF. Probiotics for treating infectious diarrhea. Cochrane Database of Systematic Reviews 2003;2.
[25] Guandalini S, Pensabene L, Zikri MA, et al. *Lactobacillus* GG administered in oral rehydration solution to children with acute diarrhea: a multicenter European trial. J Pediatr Gastroenterol Nutr 2000;30:54–60.
[26] Shornikova AV, Isolauri E, Burkanova L, Lukovnikova S, Vesikari T. A trial in the Karelian Republic of oral rehydration and *Lactobacillus* GG for treatment of acute diarrhoea. Acta Paediatr 1997;86:460–5.
[27] Pant AR, Graham SM, Allen SJ, et al. *Lactobacillus* GG and acute diarrhoea in young children in the tropics. J Trop Pediatr 1996;42:162–5.
[28] Raza S, Graham SM, Allen SJ, Sultana S, Cuevas L, Hart CA. *Lactobacillus* GG promotes recovery from acute nonbloody diarrhea in Pakistan. Pediatr Infect Dis J 1995;14:107–11.
[29] Sepp E, Tamm E, Torm S, Lutsar I, Mikelsaar M, Salminen S. Impact of a *Lactobacillus* probiotic on the fecal microflora in children with shigellosis. Microecol Ther 1995;23: 74–80.
[30] Szajewska H, Kotowska M, Mrukowicz JZ, Armanska M, Mikolajczyk W. Efficacy of *Lactobacillus* GG in prevention of nosocomial diarrhea in infants. J Pediatr 2001;138: 361–5.
[31] Mastretta E, Longo P, Laccisaglia A, et al. Effect of *Lactobacillus* GG and breast-feeding in the prevention of rotavirus nosocomial infection. J Pediatr Gastroenterol Nutr 2002; 35:527–31.
[32] Oberhelman RA, Gilman RH, Sheen P, et al. A placebo-controlled trial of *Lactobacillus* GG to prevent diarrhea in undernourished Peruvian children. J Pediatr 1999;134:15–20.
[33] Arvola T, Laiho K, Torkkeli S, et al. Prophylactic *Lactobacillus* GG reduces antibiotic-associated diarrhea in children with respiratory infections: a randomized study. Pediatrics 1999;104:e64.
[34] Vanderhoof JA, Whitney DB, Antonson DL, Hanner TL, Lupo JV, Young RJ. *Lactobacillus* GG in the prevention of antibiotic-associated diarrhea in children. J Pediatr 1999;135: 564–8.

[35] Armuzzi A, Cremonini F, Ojetti V, et al. Effect of *Lactobacillus* GG supplementation on antibiotic-associated gastrointestinal side effects during *Helicobacter pylori* eradication therapy: a pilot study. Digestion 2001;63:1–7.
[36] Cremonini F, Di Caro S, Covino M, et al. Effect of different probiotic preparations on anti-*Helicobacter pylori* therapy-related side effects: a parallel group, triple blind, placebo-controlled study. Am J Gastroenterol 2002;97:2744–9.
[37] Armuzzi A, Cremonini F, Bartolozzi F, et al. The effect of oral administration of *Lactobacillus* GG on antibiotic-associated gastrointestinal side-effects during *Helicobacter pylori* eradication therapy. Aliment Pharmacol Ther 2001;15:163–9.
[38] Siitonen S, Vapaatalo H, Salminen S, et al. Effect of *Lactobacillus* GG yoghurt in prevention of antibiotic associated diarrhoea. Ann Med 1990;22:57–9.
[39] Kalliomaki M, Salminen S, Arvilommi H, Kero P, Koskinen P, Isolauri E. Probiotics in primary prevention of atopic disease: a randomised placebo-controlled trial. Lancet 2001;357: 1076–9.
[40] Rautava S, Kalliomaki M, Isolauri E. Probiotics during pregnancy and breast-feeding might confer immunomodulatory protection against atopic disease in the infant. J Allergy Clin Immunol 2002;109:119–21.
[41] Kalliomaki M, Salminen S, Poussa T, Arvilommi H, Isolauri E. Probiotics and prevention of atopic disease: 4-year follow-up of a randomised placebo-controlled trial. Lancet 2003; 361:1869–71.
[42] Isolauri E, Arvola T, Sutas Y, Moilanen E, Salminen S. Probiotics in the management of atopic eczema. Clin Exp Allergy 2000;30:1604–10.
[43] Majamaa H, Isolauri E. Probiotics: a novel approach in the management of food allergy. J Allergy Clin Immunol 1997;99:179–85.
[44] Oksanen PJ, Salminen S, Saxelin M, et al. Prevention of travellers' diarrhoea by *Lactobacillus* GG. Ann Med 1990;22:53–6.
[45] Hilton E, Kolakowski P, Singer C, Smith M. Efficacy of *Lactobacillus* GG as a diarrheal preventive in travelers. J Travel Med 1997;4:41–3.
[46] Hatakka K, Savilahti E, Ponka A, et al. Effect of long term consumption of probiotic milk on infections in children attending day care centres: double blind, randomised trial. BMJ 2001;322:1327.
[47] Nase L, Hatakka K, Savilahti E, et al. Effect of long-term consumption of a probiotic bacterium, *Lactobacillus rhamnosus* GG, in milk on dental caries and caries risk in children. Caries Res 2001;35:412–20.
[48] Gluck U, Gebbers JO. Ingested probiotics reduce nasal colonization with pathogenic bacteria (*Staphylococcus aureus, Streptococcus pneumoniae,* and beta-hemolytic streptococci). Am J Clin Nutr 2003;77:517–20.
[49] Burke DA, Axon AT. Ulcerative colitis and *Escherichia coli* with adhesive properties. J Clin Pathol 1987;40:782–6.
[50] Favier C, Neut C, Mizon C, Cortot A, Colombel JF, Mizon J. Fecal beta-D-galactosidase production and bifidobacteria are decreased in Crohn's disease. Dig Dis Sci 1997; 42:817–22.
[51] Gupta P, Andrew H, Kirschner BS, Guandalini S. Is *Lactobacillus* GG helpful in children with Crohn's disease? Results of a preliminary, open-label study. J Pediatr Gastroenterol Nutr 2000;31:453–7.
[52] Kuisma J, Mentula S, Jarvinen H, Kahri A, Saxelin M, Farkkila M. Effect of *Lactobacillus rhamnosus* GG on ileal pouch inflammation and microbial flora. Aliment Pharmacol Ther 2003;17:509–15.
[53] Gosselink MP, Schouten WR, van Lieshout LM, Hop WC, Laman JD, Ruseler-van Embden JG. Delay of the first onset of pouchitis by oral intake of the probiotic strain *Lactobacillus rhamnosus* GG. Dis Colon Rectum 2004;47:876–84.

[54] Malin M, Suomalainen H, Saxelin M, Isolauri E. Promotion of IgA immune response in patients with Crohn's disease by oral bacteriotherapy with *Lactobacillus* GG. Ann Nutr Metab 1996;40:137–45.
[55] Prantera C, Scribano ML, Falasco G, Andreoli A, Luzi C. Ineffectiveness of probiotics in preventing recurrence after curative resection for Crohn's disease: a randomised controlled trial with *Lactobacillus* GG. Gut 2002;51:405–9.
[56] Bruzzese E, Raia V, Gaudiello G, et al. Intestinal inflammation is a frequent feature of cystic fibrosis and is reduced by probiotic administration. Aliment Pharmacol Therap 1920; 7:813–9.
[57] Biller JA, Katz AJ, Flores AF, Buie TM, Gorbach SL. Treatment of recurrent *Clostridium difficile* colitis with *Lactobacillus* GG. J Pediatr Gastroenterol Nutr 1995;21:224–6.
[58] Gorbach SL, Chang TW, Goldin B. Successful treatment of relapsing *Clostridium difficile* colitis with *Lactobacillus* GG. Lancet 1987;2:1519.
[59] Bennett RG, Gorbach SL, Goldin BR, et al. Treatment of relapsing *Clostridium difficile* diarrhea with *Lactobacillus* GG. Nutr Today 1996;31:35S–8S.
[60] Peltonen R, Kjeldsen-Kragh J, Haugen M, et al. Changes of faecal flora in rheumatoid arthritis during fasting and one-year vegetarian diet. Br J Rheumatol 1994;33:638–43.
[61] Eerola E, Mottonen T, Hannonen P, et al. Intestinal flora in early rheumatoid arthritis. Br J Rheumatol 1994;33:1030–8.
[62] Baharav E, Mor F, Halpern M, Weinberger A. *Lactobacillus* GG bacteria ameliorate arthritis in Lewis rats. J Nutr 2004;134:1964–9.
[63] Hatakka K, Martio J, Korpela M, et al. Effects of probiotic therapy on the activity and activation of mild rheumatoid arthritis—a pilot study. Scand J Rheumatol 2003;32:211–5.
[64] Mertz HR. Irritable bowel syndrome. N Engl J Med 2003;349:2136–46.
[65] Chadwick VS, Chen W, Shu D, et al. Activation of the mucosal immune system in irritable bowel syndrome. Gastroenterology 2002;122:1778–83.
[66] Gonsalkorale WM, Perrey C, Pravica V, Whorwell PJ, Hutchinson IV. Interleukin 10 genotypes in irritable bowel syndrome: evidence for an inflammatory component? Gut 2003;52:91–3.
[67] Balsari A, Ceccarelli A, Dubini F, Fesce E, Poli G. The fecal microbial population in the irritable bowel syndrome. Microbiologica 1982;5:185–94.
[68] O'Sullivan MA, O'Morain CA. Bacterial supplementation in the irritable bowel syndrome: a randomised double-blind placebo-controlled crossover study. Dig Liver Dis 2000; 32:294–301.
[69] Goldin BR, Gualtieri LJ, Moore RP. The effect of *Lactobacillus* GG on the initiation and promotion of DMH-induced intestinal tumors in the rat. Nutr Cancer 1996;25:197–204.
[70] Lim BK, Mahendran R, Lee YK, Bay BH. Chemopreventive effect of *Lactobacillus rhamnosus* on growth of a subcutaneously implanted bladder cancer cell line in the mouse. Jpn J Cancer Res 2002;93:36–41.
[71] Lahtinen SJ, Haskard CA, Ouwehand AC, Salminen SJ, Ahokas JT. Binding of aflatoxin B1 to cell wall components of *Lactobacillus rhamnosus* strain GG. Food Addit Contam 2004;21:158–64.
[72] Haskard C, Binnion C, Ahokas J. Factors affecting the sequestration of aflatoxin by *Lactobacillus rhamnosus* strain GG. Chem Biol Interact 2000;128:39–49.
[73] Grun M, Liehr H, Rasenack U. Significance of endotoxaemia in experimental ''galactosamine-hepatitis'' in the rat. Acta Hepatogastroenterol (Stuttg) 1977;24:64–81.
[74] Camara DS, Caruana JA Jr, Schwartz KA, Montes M, Nolan JP. D-Galactosamine liver injury: absorption of endotoxin and protective effect of small bowel and colon resection in rabbits. Proc Soc Exp Biol Med 1983;172:255–9.
[75] Wang B, Egashira Y, Ohta T, Sanada H. Effect of indigestible oligosaccharides on the hepatotoxic action of D-galactosamine in rats. Biosci Biotechnol Biochem 1998;62: 1504–9.

[76] Nanji AA, Khettry U, Sadrzadeh SM. *Lactobacillus* feeding reduces endotoxemia and severity of experimental alcoholic liver (disease). Proc Soc Exp Biol Med 2005;3:243–7.
[77] Bauer TM, Fernandez J, Navasa M, Vila J, Rodes J. Failure of *Lactobacillus* spp. to prevent bacterial translocation in a rat model of experimental cirrhosis. J Hepatol 2002;36:501–6.
[78] Dong MY, Chang TW, Gorbach SL. Effects of feeding *Lactobacillus* GG on lethal irradiation in mice. Diagn Microbiol Infect Dis 1987;7:1–7.
[79] Tabuchi M, Ozaki M, Tamura A, et al. Antidiabetic effect of *Lactobacillus* GG in streptozotocin-induced diabetic rats. Bioscience. Biotechnol Biochem 2003;67:1421–4.
[80] Cooke KR, Olkiewicz K, Erickson N, Ferrara JL. The role of endotoxin and the innate immune response in the pathophysiology of acute graft versus host disease. J Endotoxin Res 2002;8:441–8.
[81] Beelen DW, Elmaagacli A, Muller KD, Hirche H, Schaefer UW. Influence of intestinal bacterial decontamination using metronidazole and ciprofloxacin or ciprofloxacin alone on the development of acute graft-versus-host disease after marrow transplantation in patients with hematologic malignancies: final results and long-term follow-up of an open-label prospective randomized trial. Blood 1999;93:3267–75.
[82] Gerbitz A, Schultz M, Wilke A, et al. Probiotic effects on experimental graft-versus-host disease: let them eat yogurt. Blood 2004;103:4365–7.

Gastroenterol Clin N Am 34 (2005) 499–513

GASTROENTEROLOGY CLINICS
OF NORTH AMERICA

ELSEVIER
SAUNDERS

VSL#3: An Analysis of Basic and Clinical Contributions in Probiotic Therapeutics

Paolo Gionchetti, MD*, Karen M. Lammers, PhD, Fernando Rizzello, MD, Massimo Campieri, MD

Department of Internal Medicine and Gastroenterology, Policlinic Sant'Orsola, University of Bologna, Via Massarenti 9, 40138 Bologna, Italy

Microflora represent a heterogeneous mixture of microorganisms that are distributed throughout the gastrointestinal (GI) tract, but the highest concentrations of bacteria are present in the large intestine. Circa 400 bacterial species constitute the colonic flora. Compared with the huge concentration of bacteria in the GI tract, the stomach contains a relatively small number of bacteria ($<10^3$ colony forming units [CFU]/mL) because of low gastric pH. The isolated bacteria are predominantly gram-positive and aerobic. Streptococci, staphylococci, lactobacilli and various fungi are the most commonly isolated species. The concentration of microbes increases from about 10^4 CFU/mL in the proximal small intestine to a concentration of 10^5 to 10^9 CFU/mL in the distal ileum. Although the bacterial composition in the proximal ileum is quite similar to that of the stomach, there is an increasing predominance of gram-negative bacteria in the distal ileum. Coliforms and anaerobic bacteria such as *Bacteroides*, *Bifidobacterium*, *Clostridium,* and *Fusobacterium* can be isolated in substantial concentrations. Within the colon, the total number of microorganisms increases to 10^{10} to 10^{12} CFU/mL, and anaerobes outnumber aerobes by a ratio of 1000:1. The predominant isolates are *Bacteroides*, *Bifidobacterium,* and *Eubacterium*, whereas anaerobic gram-positive cocci, clostridia, enterococci, and some species of *Enterobacteriaceae* are present too [1,2].

A substantial part of the microflora still needs to be described; analysis and characterization of the hundreds of bacterial species within the ecosystem counter difficulties related to specific culture conditions such as lack of specificity of culture media, different growth properties of species in selective media, nonviability or noncultivability because of their special requirements for anaerobiosis, or complex nutritional interactions that can occur between the inhabitants of bacterial communities and which may be difficult to achieve in laboratory culture media [3].

*Corresponding author. *E-mail address*: paolo@med.unibo.it (P. Gionchetti).

0889-8553/05/$ – see front matter
doi:10.1016/j.gtc.2005.05.003

gastro.theclinics.com

HOST–MICROBIOTA INTERACTIONS

A balanced interplay exists between the intestinal microflora and the host. To maintain this balanced condition, the host has acquired protective defensive mechanisms including gastric acid, peristalsis and desquamation, the epithelium, the secretion of antimicrobial peptides, and the mucosal immune system. In its turn, the presence of the indigenous flora is crucial for digestion of nutrients and production of vitamins, the development of a normal intestinal morphology, and maturation of the mucosal immune system [1,2].

Gastric acid is protective against infection with pathogenic microorganisms and influences indirectly bacterial titer and composition in the upper small intestine. Intestinal motility is the major defense against bacterial overgrowth. Bacteria are removed efficiently by peristalsis. The epithelium lines the intestinal mucosa and is in direct contact with the intestinal lumen. The intestinal epithelium consists of a monolayer of columnar, polarized cells firmly connected by tight junctions (TJs), dynamic structures that represent the major barrier within the paracellular pathway and regulate in the paracellular permeability a rapid and coordinated way [4]. At the luminal side, enterocytes express proteins necessary for absorption and digestion of nutrients, water, and electrolytes. The basolateral epithelial membranes express ion transporters, which regulate water and electrolyte transport. Of importance in maintaining epithelial barrier integrity are, in addition to the TJ proteins, factors that are released by the epithelium and interfere with bacterial attachment. These protective factors include mucin, secretory IgA, chloride, the recently described trefoil peptides, and antimicrobial defensins [5–7]. Commensal bacteria compete for binding sites in the mucus layer to prevent attachment of pathogenic bacteria. Production of antimicrobial substances, such as bacteriocins and hydrogen peroxide, and introduction of a low pH create an environment unfavorable and inhibitory for certain enteric pathogens. Additionally, the microflora provide nutrients for the host through involvement in the production of vitamins and in metabolizing substrates. Common end products of microbial fermentation are the short chain fatty acids (eg, acetic, propionic, and butyric acids). Short chain fatty acids are a major source of energy for epithelial cells and play a role in maintaining integrity of the colonic epithelial layer and in the regulating cell growth and differentiation of epithelial cells [1,2]. Given its localization, the intestinal epithelium probably provides the first level of regulatory control, either through its ability to take up antigens, or through its ability to respond to luminal antigens by producing mediators that can influence underlying mucosal immune cells [8]. Bacterial products are recognized by Toll-like receptors (TLR) expressed at the apical side of epithelial cells [9] and subsequently transported in luminal-to-serosal direction to the basolateral surface [10], which is in close proximity to cell populations in the lamina propria. Undifferentiated epithelial cells located in the crypts can produce interleukin (IL)-1β [11]. Upon stimulation with various stimuli, both primary intestinal epithelial cells and cell lines [12] can produce interleukin (IL)-8 (CXCL8), a potent chemoattractant cytokine for neutrophils

[13], IL-6 [14], and many chemokines [15,16]. In normal condition, the interaction between the luminal antigenic load, immune cells, and nonimmune cells favors induction of tolerance to ensure maintenance of homeostasis and preservation of the integrity of the intestinal mucosa. In this way, the intestinal immune system retains the ability to distinguish between commensal bacteria or dietary proteins and pathogens and to maintain tolerance or to exert a strong protective immune response to prevent invasion of pathogens [8].

BACTERIA IN INFLAMMATORY BOWEL DISEASE

Clinical and experimental observations suggest that luminal bacteria and their products are implicated in the pathogenesis of inflammatory bowel disease (IBD) in a genetically predisposed individual. The distal ileum and the colon are the areas with the highest bacterial concentrations, and these sites most frequently are affected in IBD. Similarly, pouchitis appears to be associated with bacterial overgrowth and dysbiosis. Patients with pouchitis may be treated effectively with antibiotics. Enteric bacteria or their phlogistic products have been found within the inflamed mucosa of patients with Crohn's disease. Lowering the bacterial concentration by antibiotics or diverting the fecal stream decreases disease activity in Crohn's disease, with recurrence of disease occurring after restoration of fecal stream [17,18]. Recent studies have shown the ability of luminal contents, presumably dominated by bacteria or their products, to trigger postoperative recurrence of Crohn's disease in the neoterminal ileum within few days [19]. Macpherson and colleagues showed that total mucosal IgG titers were high in patients who had IBD and that the antibodies were directed against bacterial cytoplasmic proteins rather than membrane proteins [20]. Compelling evidence is derived from animal models of colitis. Despite the great diversity in genetic defects and immunopathology in the existing transgenic and knockout mutant murine models of colitis, the presence of a normal enteric flora is a prerequisite for the induction of inflammation in all of these models [21].

For years, researchers have tried in vain to identify a specific pathogen as cause for chronic intestinal inflammatory disorders, yet it cannot be excluded that one or more bacterial agent(s) is/are responsible. Considering a possible implication of a pathogen in IBD, not yet being discovered because of technical limitations, it was hypothesized that modulating abnormal microflora from these patients by introducing high titers of protective bacteria might overwhelm the aggressive strain(s) and inhibit deleterious effects. On this basis, probiotic treatment was proposed as a therapeutic approach [22].

PROBIOTICS

The use of lactic acid bacteria in food has a long history. Fermented food has been consumed for centuries. The first to propose the use of probiotics for the purpose of health maintenance and disease prevention, however, was Elie Metchnikoff (1845 to 1916), the Russian Nobel Prize Laureate, who at the beginning of the last century suggested that high concentrations of lactobacilli in the intestinal flora were important for health and longevity in people [23]. He considered the gut microflora as a source of autointoxication that might have repercussions on the aging process. In particular, bacteria that degrade proteins (putrefaction) were harmful for the host because of the release of ammonia, amines, and indole, which are toxic to the nervous and vascular systems and were absorbed by the bowel and subsequently entered the systemic circulation. To prevent this condition of autointoxication, Metchnikoff proposed to replace and to diminish the concentration of harmful bacteria by ingesting bacterial populations that fermented carbohydrates and had little proteolytic activity [23]. At the same time, Tissier showed that bifidobacteria were the predominant intestinal microbial flora in breastfed infants, and it was speculated that infant diarrhea could be treated by the administering large doses of bifidobacteria [24]. The term probiotic first was introduced in 1965 by Lilly and Stillwell to describe any substance or organism that contributes to intestinal microbial balance [25]. This definition was modified by Fuller, who stressed the importance of a live microbial feed supplement to beneficially affect the host by improving the intestinal microbial balance. More recently, probiotics have been defined as living organisms, which upon ingestion in certain numbers, exert health benefits beyond inherent basic nutrition [26], emphasizing the importance of maintaining a sufficient population of live microorganisms in the gut, and further implying other effects such as improvement in microbial balance and immunomodulation [27].

Most probiotics belong to a group of lactic acid bacteria (lactobacilli, streptococci) and bifidobacteria that are important components of the natural microflora. Members of the genus *Lactobacillus* are nonsporing, gram-positive, rod-shaped bacteria. Lactobacilli are facultative anaerobic bacteria that produce large amounts of lactic acid by fermenting glucose. Lactobacilli constitute only 1% of the fecal microflora. The composition of the fecal microflora reflects that of the distal colon, while facultative anaerobes have been shown to be more numerous in the cecal contents compared with the feces, where obligate anaerobes predominate. It therefore should not be excluded that the low percentage of facultative anaerobic bacteria detected in feces represents remnants of much larger populations inhabiting the proximal colon. Members of the genus *Bifidobacterium* are nonsporing, nonmotile, gram-positive bacteria. Dependent on culture conditions, they may have a rod-shaped or branching morphology. They are obligate anaerobic bacteria and produce large amounts of acetic acid and lactic acid from the fermentation of glucose. Dependent on the techniques used, bifidobacteria represent up to 10% of the fecal microflora in adult humans, and up to 91% of the microflora of breastfed infants [3]. Other probiotics also include some nonpathogenic strains of *Escherichia coli* and other microorganisms such as the yeast *Saccharomyces boulardii.*

For clinical application, probiotic strains need to have certain characteristics. Because some health-promoting effects may be species-specific, the probiotic strains must be of human origin. They must be normal inhabitants of the intestinal tract and need to be resistant to the acid of the stomach, digestive enzymes, and bile salts. They must be able to exert metabolic activity within intestinal lumen, where, ideally, they should survive, but not persist long-term. The strains should be antagonistic against pathogenic bacteria, and, obviously, they must be safe for human use and should maintain their viability and beneficial properties during manufacturing processes [28].

Different probiotic preparations are being investigated for inflammatory and functional bowel disorders. This article focuses on one of these preparations, VSL#3. The probiotic preparation VSL#3 is characterized by a very high bacterial concentration of 300 billion/g probiotic bacteria consisting of eight strains of viable lyophilized bacteria; four strains of *Lactobacillus* (*L casei, L plantarum, L acidophilus, L delbrueckii* subspecies bulgaricus), three strains of *Bifidobacterium* (*B longum, B breve, B infantis*) and one strain of *Streptococcus salivarius* subspecies *thermophilus* (designated throughout the text as *S thermophilus*). Rationale for the use of a cocktail containing large numbers of different strains has been the concept of high efficacy through a synergistic action of the different strains in the mixture.

VSL#3 IN EXPERIMENTAL STUDIES

The probiotic concept remains controversial, because the precise mechanisms by which probiotic microorganisms exert their protective effects in vivo have not been defined well [29]. Many researchers are attempting to unravel the precise mechanisms by which probiotic bacteria and their metabolic products (short-chain fatty acids, vitamins) exert their beneficial effects. Several mechanisms have been proposed to account for probiotic action. These include:

- The antagonistic activity against pathogenic bacteria by competing for binding sites
- Inhibition of bacterial translocation and reinforcement of barrier function
- Stimulation of mucosal defense both at the level of immune and epithelial function by increasing the production of secretory IgA, by stimulation of protective epithelial substances (such as mucin), by regulation of mucosal cytokine production, and by production of nutrients of special importance to the intestine such as short-chain fatty acids and vitamins

Mucosal host defense

Rachmilewitz and colleagues have studied the probiotic preparation VSL#3, a highly concentrated cocktail of eight different probiotic strains, in a model of iodoacetamide-induced colitis in rats. The preparation appeared to be effective in the attenuation of inflammation and determined a decrease in NOS (nitric oxide synthase)- and MPO (myeloperoxidase)- and activity, which are indicative for a decrease in neutrophil infiltration [30]. The efficacy of VSL#3 probiotic bacteria was studied further in the IL-10 gene-deficient mice. These mice develop spontaneous colitis with features similar to inflammation observed in patients with Crohn's disease. Histological damage begins within

4 weeks and reaches a plateau after 8 weeks. Within 2 weeks of age and before the appearance of histological injury, there appears to be a primary alteration in bacterial colonization, with an increase in mucosal adhered and translocated bacteria. These alterations reach a steady state approximately 1 month before maximal histological injury and persist for months. The same compound was used as a primary therapy in colitis in the IL-10 gene-deficient mouse model. In this study, a significant improvement of inflammation was reported together with a reduction in mucosal levels of the proinflammatory cytokines tumor necrosis factor α (TNF-α) and interferon-γ (IFN-γ), and with normalization of the colonic barrier integrity. Furthermore, the bacteria within the VSL#3 preparation increased monolayer resistance and decreased mannitol permeability in T84 epithelial cells, which is indicative for an enhancement of barrier function. Moreover, bacteria within this compound were able to reduce pathogenic bacterial invasion in T84 epithelial cells [31].

Probiotic therapy with VSL#3 in (IPAA) patients with ileal-pouch anal anastomosis within the first year after IPAA surgery [32] reduced the mucosal gene expression of the proinflammatory cytokines IL-1β, IFN-γ, and IL-8. Morphological data revealed that less polymorphonuclear cells were present in the mucosal tissue from patients within the probiotic group than in mucosal tissue from patients within the placebo group. These data suggest that probiotic therapy affects IL-8 production at molecular level and limits the influx of polymorphonuclear cells. The decreased IFN-γ and IL-1β gene expression may have a role in regulating the inflammatory response by downregulation of T cell activation or induction of regulatory T cells, reinforcement of the intestinal barrier function, and a tight control of the proinflammatory cytokine IL-1β, respectively [33]. Significantly elevated mucosal IL-10 levels, in addition to a significant decrease of mucosal levels of IFN-γ, IL-1 and matrix metalloproteinase activity, were found in chronic relapsing pouchitis patients who were treated with probiotics immediately after the patients had entered in remission by antibiotic therapy [34].

Two studies examined the effects of probiotic bacteria on IL-8 by intestinal epithelial cells and found a species specificity of the induced effects [35,36]. Authors found that whole bacteria and bacterial cell fractions of VSL#3 probiotic bacteria did not induce IL-8 production by intestinal epithelial cells, whereas the whole bacteria and bacterial cell fractions of the strain *E coli* Nissle 1917 induced IL-8 production. VSL#3 probiotic bacteria dose-dependently inhibited IL-8 production that was produced by *E coli* Nissle and by the pathogen *Salmonella dublin*. IL-8 production was higher in response to bacterial cell extracts.

In a previous article on VSL#3 probiotic bacteria, the authors expected a proteinaceous soluble factor to be responsible for the effects they observed on immunity and epithelial barrier function [31]. Otte and Podolsky described their efforts to characterize this soluble factor and found that although bacterial DNA might account in part for the observed protective or immunomodulatory effects, a heat-sensitive protein with a molecular weight probably exceeding

50 kDa likely was responsible for these effects [36]. A recent study showed that VSL#3 produced soluble factors that inhibited the chymotrypsin-like activity of the proteasome in gut epithelial cells. In addition, probiotic bacteria inhibited the proinflammatory nuclear factor-kappaB pathway and induced the expression of protective heat shock proteins [37].

The immune effects of probiotic bacterial DNA were investigated in several experimental models, as various reports showed that CpG motifs, short sequences of unmethylated bacterial DNA, have immunostimulatory properties [38,39]. Rachmilewitz and colleagues [40] showed that probiotic bacterial DNA has protective effects and that signaling by means of TLR9, the receptor required for recognition of bacterial DNA [41], mediated the observed anti-inflammatory effects; administration of probiotic DNA attenuated the severity of DSS-induced colitis, and the beneficial effects of the VSL#3 probiotics were abolished in the presence of chloroquine, an inhibitor of TLR9 signaling, or in TLR9 gene-deficient mice. Madsen and colleagues [42] showed that VSL#3 DNA reduced inflammatory responses in vivo and in vitro, both at systemic and local level. IL10 gene-deficient mice were fed probiotic DNA or were injected intraperitoneally with *E coli* DNA and VSL#3 DNA. Probiotic bacterial DNA inhibited mucosal tissue IFN-γ and TNF-α secretion and improved histologic grading. The systemic TNF-α release that was induced by intraperitoneal injection of *E coli* DNA was attenuated by VSL#3 DNA. Furthermore, stimulation of HT29 epithelial cell monolayers with probiotic bacterial DNA in the presence of TNF-α inhibited the TNF-α-induced IL-8 secretion by HT29 cells. In addition, activation of p38 MAPK was reduced; NFκB activation was delayed; IκB levels were stabilized, and proteasome function was inhibited. In another study, the pure genomic DNA derived from each VSL#3 probiotic bacterial strain was applied in peripheral blood mononuclear cells (PBMC) cultures [43]. This study showed that stimulation with probiotic bacterial DNA induced strain-specific IL-1 and IL-10 production by PBMC. In particular, DNA derived from bifidobacteria was a potent IL-10 inducer by PBMC. To evaluate whether bacterial DNA released from ingested probiotic bacteria could have a modulatory effect on immune cells, total bacterial DNA was extracted from feces from volunteers before probiotic ingestion (T_0) and after 2 weeks of probiotic ingestion (T_1). PBMCs were stimulated with these bacterial DNAs derived from feces, and cytokine production was measured. Total fecal bacterial DNA after modulation of the microflora by probiotic ingestion increased IL-10 and decreased IL-1 production, demonstrating that DNA of the probiotic bacteria ingested is capable of exerting an immune effect. Further stimulation experiments with probiotic bacterial DNA were performed in a whole blood assay to assess dendritic cell phenotype and function. Probiotic bacterial DNA induced high IL-10 production, but no IL-12 and IFN-α by dendritic cells (DC). Preliminary data suggest that the bacterial DNA-induced IL-10 production may be dependent of TLR9, at least partially, as blocking of this receptor inhibited IL-10 production [44].

Two other studies evaluated the immune effects of VSL#3 in DC. Hart and colleagues [45] studied the effect of cell wall and cell membrane fractions of single VSL#3 bacterial strains and the mixture in blood dendritic cells and human colonic lamina propria dendritic cells. In addition, the subsequent T cell response after exposure to bacterial cell fractions was assessed. VSL#3 bacterial strains had strain-specific, immunomodulatory effects on DC; after stimulation with VSL#3 probiotic bacterial fractions, DC from blood and intestinal tissue produced significant IL-10, and the generation of Th1 cells was inhibited. In particular, bifidobacteria exerted strong anti-inflammatory effects, that is high IL-10 production and decreased expression of the costimulatory molecule CD80 by DC, and decreased IFN-γ production by T cells. When coincubated with lipopolysaccharide (LPS), VSL#3 bacterial fractions decreased the LPS-induced production of IL-12 while IL-10 production was maintained. Drakes and colleagues [46] showed substantial IL-10, but not IL-12, production by murine bone marrow-derived DC in response to stimulation with the VSL#3 mixture.

Antagonistic activity against pathogenic bacteria, inhibition of bacterial translocation and reinforcement of barrier function

Maintaining epithelial barrier function is necessary for preserving mucosal integrity. Altered TJ structure in ulcerative colitis results in impaired barrier function [47]. Increased permeability in IBD is observed during active phase of the disease. Therefore, reinforcement of the barrier function, together with their immune modulatory and metabolic properties, may be central in the mechanism of action of probiotic bacteria. The exact mechanisms by which probiotics can influence barrier function remain to be elucidated. It is known that certain lactobacilli adhere to mucosal surfaces, inhibit the attachment of pathogenic bacteria, and enhance the secretion of mucins [48–50]. These properties may be instrumental in improving the mucosal barrier function and decreasing the permeability to macromolecules and toxins. It recently was shown that VSL#3 probiotic bacteria, in addition to decreasing proinflammatory cytokines [31,34–36,42–45], reinforce the barrier function by secreting soluble factors that enhance the barrier integrity and by regulating TJs [31,36]. TJs are dynamic structures, represent the major barrier within the paracellular pathway, and regulate in a rapid and coordinated way the paracellular permeability. Many pathogenic bacteria modulate intestinal permeability by altering TJs [4]. To study the effects of probiotics and pathogenic bacteria on epithelial barrier function, changes in T84 epithelial cell culture integrity and confluence were measured by means of transepithelial resistance (TER), and changes in paracellular permeability were measured by evaluating zonula occludens-1 (ZO-1) distribution. Infection of epithelial cells with *S dublin* resulted in significant decrease in TER and a distinct distribution of ZO-1 compared with noninfected T84 cells. Coincubation with probiotics prevented *S dublin*-induced alteration of distribution of the TJ protein ZO-1 and attenuated *S dublin*-induced enhanced permeability by increasing the transepithelial resistance and by reducing the mannitol flux [31,36]. In this assay, probiotic bacteria-conditioned medium was less effective than live bacteria. Probiotic bacteria were also able to induce mucin gene expression [36].

VSL#3 IN CLINICAL STUDIES

VSL#3 and ulcerative colitis

In a first pilot study, the probiotic preparation VSL#3 was used as maintenance treatment in patients with ulcerative colitis in remission allergic or intolerant to sulphasalazine and mesalazine to verify the impact of this preparation on the fecal flora. Twenty patients received VSL#3 6 g/d for 12 months. The patients were assessed clinically every 2 months. Endoscopic evaluation was performed at the beginning of the study, at 6 and 12 months, or, in case of relapse. Stool culture and determination of fecal pH also were performed at different intervals during the study, and the last stool culture was determined 15 days after the end of the treatment. Fecal concentrations of lactobacilli, bifidobacteria and *S thermophilus* increased significantly in all patients. This increase became evident already after 20 days of VSL#3 ingestion and persisted through the treatment period. Within 15 days after the end of the treatment, bacterial concentrations of lactobacilli, bifidobacteria and *S thermophilus* turned to basal levels, whereas no modification of fecal concentrations of *Bacteroides*, enterococci, coliforms, *Clostridia* and total anaerobes and aerobes was observed. Fecal pH was reduced significantly during the treatment, and 15 of the 20 patients (75%) remained in remission [51]. More recently, in another uncontrolled pilot study, 30 patients with active mild-to-moderate ulcerative colitis were administered high-dose VSL#3 12 g (3600 billion bacteria)/d. Using a nucleic acid-based method (PCR/ Denaturing Gradient Gel Electrophoresis [DGGE]), the bacterial components of the VSL#3 preparation and the composition of the biopsy-associated microbiota in a subset of patients before and after the administration of VSL#3 were evaluated. Remission was achieved in 63% (n = 19) and response in 23% (n = 7) of the patients. All eight components of VSL#3 could be detected by PCR/DGGE, and at least some bacterial species present in the probiotic preparation could be detected in the biopsies [52].

VSL#3 and pouchitis

Subsequently, the efficacy of this probiotic preparation versus placebo in the maintenance treatment of chronic pouchitis was evaluated in a double-blind study. Forty patients who obtained clinical and endoscopic remission after 1 month of combined antibiotic treatment (ciprofloxacin 1 g/d and rifaximin 2 g/d), were randomized to receive VSL#3 6 g (1800 billion bacteria)/d or an identical appearing placebo for 9 months. Patients were assessed clinically every month, and endoscopic and histological assessments were performed at entry and then every 2 months. Relapse of pouchitis was defined as an increase of at least two points in the clinical portion of Pouchitis Disease Activity Index (PDAI) and was confirmed endoscopically and histologically. For the determination of fecal concentrations of lactobacilli, bifidobacteria, and *S thermophilus*, stool culture was performed before and after antibiotic treatment and subsequently every month during maintenance treatment. Of the 20 patients within the placebo group, all had a relapse in the follow-up period, eight within 2 months, seven within 3 months, and five within 4 months. In contrast, 17 of the 20 patients (85%) treated with VSL#3 remained in remission after 9 months ($P < .001$). All 17 patients had a relapse within 4 months after suspension of the active treatment, and the mean duration of remission was 2 months. The mean total PDAI score of the 20 relapsed patients treated with placebo was 12 (range 8 to 18); median stool frequency was 9 (range 7 to 11). In the group treated with VSL#3, the three patients who had relapses during the 9 months of follow-up had a median total PDAI score of 11 (range 9 to 17). Median stool frequency in these patients was 8 (range 6 to 11) at the time of relapse. The 17 patients who remained in remission had a median total PDAI score of 0. The median stool frequency in these patients did not increase significantly compared with that obtained after antibiotic treatment (four [range three to six]). Fecal concentrations of lactobacilli, bifidobacteria and *S thermophilus* were increased significantly within 1 month after starting VSL#3 treatment and remained stable throughout the study. The increase in the levels of lactobacilli, bifidobacteria, and *S thermophilus* did not affect the concentration of the other bacterial groups determined, including *Bacteroides*, coliforms, clostridia, and enterococci. These data suggest that the effect was not mediated by suppression of endogenous luminal bacteria [53].

A randomized, placebo-controlled study aimed to investigate the effectiveness of a single daily dose of VSL#3 in maintaining antibiotic-induced remission (obtained after 1 month of treatment with a combination of metronidazole 800 mg/d and ciprofloxacin 1 g/d) in patients with refractory or recurrent pouchitis. The study period was 1 year, and clinical, endoscopic, and histological assessments were made before, at 2 months, and at the end of the study period. In parallel, an assessment of health-related quality of life was performed by Inflammatory Bowel Disease Questionnaire (IBDQ). The study demonstrated and confirmed the clinical efficacy of the probiotic preparation as maintenance treatment in patients who had chronic pouchitis, with a remission rate of 85% in the probiotic-treated group versus 6% in the placebo group. A significant increase in IBDQ score was associated with the clinical efficacy of bacteriotherapy [54].

Encouraging results were obtained from a recent pilot study that aimed to evaluate whether a high dose of VSL#3 12 g (3600 billion bacteria)/d applied for 1 month can be useful as primary therapy in patients with mild active pouchitis and improve quality of life. Remission was achieved in 13 of the 19 patients treated (63%), and the median IBDQ score was increased significantly [55].

VSL#3 and the prevention of pouchitis

A double-blind, placebo-controlled trial was performed to evaluate the efficacy of VSL#3 in preventing pouchitis onset in 40 patients operated of ileal pouch-anal anastomosis for ulcerative colitis. Within 1 week after ileostomy closure, patients were randomized to receive VSL#3 3 g (900 billion bacteria)/d or an identical placebo for 12 months; patients were assessed clinically, endoscopically and histologically at 1, 3, 6, 9, and 12 months according to PDAI. Patients treated with the probiotic compound had a significantly lower incidence of acute pouchitis (10%) compared with those treated with placebo (40%) during the first year after ileostomy closure ($P < .05$). Treatment with VSL#3 determined a significant and progressive reduction of stool frequency and improved IBDQ score [33].

VSL#3 and Crohn's disease

A single-blind study compared a combined antibiotic–probiotic treatment with mesalazine treatment on the prevention of postoperative recurrence of Crohn's disease. Forty patients who underwent surgery were randomized within 1 week after operation to receive either a high dose of rifaximin, a nonabsorbable antibiotic, for 3 months followed by VSL#3 6 g/d for 9 months or mesalazine 4 g/d for 12 months. Patients were assessed clinically and endoscopically at 3 and 12 months. The combined antibiotic–probiotic treatment resulted in a significant lower incidence of severe endoscopic recurrence at 3 and 12 months, with 10% recurrence in the combination group versus 40% recurrence in the mesalazine group, and 20% recurrence in the combination groups versus 40% recurrence in the mesalazine group, respectively ($P < .01$) [55].

SUMMARY

These are early days in probiotic research. The mechanism of action of probiotic bacteria remains incompletely defined, yet the information gathered over the last few years starts to depict how immunological and immunogenetic factors may influence intestinal homeostasis and how important the composition of the microflora is in the delicate and dynamic balanced interaction between host and microbiota. The data from experimental studies indicate that the VSL#3 bacterial strains have immunomodulatory effects. Attenuation of severity of disease activity by means of improvement of histologic grading has been described in various animal models. A decrease of neutrophil tissue influx and activity has been shown. Further, a diminished proinflammatory IL-1β, TNF-α, IFN-γ, IL-12, and IL-8 cytokine production and an enhanced production of the anti-inflammatory cytokine IL-10 have been reported. Coincubation of probiotic bacteria with proinflammatory stimuli or pathogenic bacteria revealed that probiotic bacteria inhibited the proinflammatory cytokine production induced by these inflammatory stimuli, and further, that probiotics prevented *Salmonella*-induced damage of the intestinal barrier integrity. Maintenance of the epithelial barrier function is necessary for preserving mucosal integrity. Altered TJ structure in ulcerative colitis results in impaired barrier function. Increased permeability in IBD is observed during the active phase of the disease. Therefore, reinforcement of the barrier function, together with immune-modulatory and metabolic properties, may be central in the mechanism of action of probiotic bacteria. Further extensive investigation on how probiotics regulate the TJ complex and influence intestinal permeability is warranted.

In vivo encouraging results have been obtained from various well-designed studies on the effect of the probiotic preparation VSL#3 in patients with ulcerative colitis and ileal pouch anal anastomosis for ulcerative colitis. The health-promoting properties of probiotic bacteria by their metabolic activity and adhesion properties underscore the importance of administering live bacteria for therapeutic purpose. The promising results of studies on probiotic bacterial DNA need further investigation to elucidate the possible importance of probiotic bacterial DNA. Given the high titers of bacteria in the gut lumen, it seems opportune consider that a substantial release of bacterial DNA caused by physiologic bacterial lysis in the intestine may contribute to immune homeostasis.

Probiotics may provide a simple and attractive way to treat IBD, and patients find the probiotic concept appealing because they are natural, safe, and well-tolerated. Highly concentrated probiotic preparations are a valid approach for preventing pouchitis onset and relapses. The encouraging results obtained in ulcerative colitis and pouchitis and the variable results obtained in Crohn's disease need to be assessed further in large double-blind trials using well-characterized probiotic preparations. For a potent clinically active probiotic preparation, most likely a combination of high probiotic bacterial concentrations and different probiotic strains are needed. Probiotics may represent a useful adjunctive treatment to conventional therapy. Attention should be paid to the selection of well-defined probiotic preparations. The viability and survival of strains within many available preparations are not documented. The benefit of one probiotic preparation does not imply the efficacy of other preparations containing different strains. Each strain exerts very different and specialized functions. A better understanding of how bacteria function in the gut ecosystem, and how bacteria interact with the host, is crucial to recognizing the potential advantages of probiotic use and could provide essential information on the pathogenesis of IBD.

References

[1] Simon GL, Gorbach SL. Intestinal flora in health and disease. Gastroenterology 1984; 86:174–93.

[2] Guarner F, Malagelada JR. Gut flora in health and disease. Lancet 2003;361:512–9.

[3] Tannock GW. Probiotics and prebiotics—where are we going? Wymondham, Norfolk, United Kingdom: Caister Academic Press; 2002. p. 1–39.

[4] Fasano A, Nataro JP. Intestinal epithelial tight junctions as targets for enteric bacteria-derived toxins. Adv Drug Deliv Rev 2004;56:795–807.

[5] Tollin M, Bergman P, Svenberg T, et al. Antimicrobial peptides in the first line defense of human colon mucosa. Peptides 2003;24:523–30.

[6] Fellermann K, Stange EF. Defensinsinnate—immunity at the epithelial frontier. Eur J Gastroenterol Hepatol 2001;13:771–6.

[7] Goke M, Podolsky DK. Regulation of the mucosal epithelial barrier. Baillieres Clin Gastroenterol 1996;10:393–405.

[8] Mowat AM. Anatomical basis of tolerance and immunity to intestinal antigens. Nat Rev Immunol 2003;3:331–41.
[9] Cario E, Brown D, McKee M, et al. Commensal-associated molecular patterns induce selective toll-like receptor-trafficking from apical membrane to cytoplasmic compartments in polarized intestinal epithelium. Am J Pathol 2002;160:165–73.
[10] Gewirtz AT, Navas TA, Lyons S, et al. Cutting edge: bacterial flagellin activates basolaterally expressed TLR5 to induce epithelial proinflammatory gene expression. J Immunol 2001;167:1882–5.
[11] Radema SA, van Deventer SJ, Cerami A. Interleukin 1 beta is expressed predominantly by enterocytes in experimental colitis. Gastroenterology 1991;100:1180–6.
[12] Yang SK, Eckmann L, Panja A, et al. Differential and regulated expression of C–X-C, C–C, and C-chemokines by human colon epithelial cells. Gastroenterology 1997;113: 1214–23.
[13] Baggiolini M. Chemokines in pathology and medicine. J Intern Med 2001;250:91–104.
[14] Sitaraman SV, Merlin D, Wang L, et al. Neutrophil-epithelial crosstalk at the intestinal lumenal surface mediated by reciprocal secretion of adenosine and IL-6. J Clin Invest 2001;107:861–9.
[15] Shibahara T, Wilcox JN, Couse T, et al. Characterization of epithelial chemoattractants for human intestinal intraepithelial lymphocytes. Gastroenterology 2001;120:60–70.
[16] Dwinell MB, Lugering N, Eckmann L, et al. Regulated production of interferon-inducible T-cell chemoattractants by human intestinal epithelial cells. Gastroenterology 2001;120: 49–59.
[17] Janowitz HD, Croen EC, Sachar DB. The role of the fecal stream in Crohn's disease: an historical and analytic review. Inflamm Bowel Dis 1998;4:29–39.
[18] Sutherland L, Singleton J, Sessions J, et al. Double blind, placebo controlled trial of metronidazole in Crohn's disease. Gut 1991;32:1071–5.
[19] D'Haens GR, Geboes K, Peeters M, et al. Early lesions of recurrent Crohn's disease caused by infusion of intestinal contents in excluded ileum. Gastroenterology 1998; 114:262–7.
[20] Macpherson A, Khoo UY, Forgacs I, et al. Mucosal antibodies in inflammatory bowel disease are directed against intestinal bacteria. Gut 1996;38:365–75.
[21] Elson CO, Sartor RB, Tennyson GS, et al. Experimental models of inflammatory bowel disease. Gastroenterology 1995;109:1344–67.
[22] Campieri M, Gionchetti P. Probiotics in inflammatory bowel disease: new insight to pathogenesis or a possible therapeutic alternative? Gastroenterology 1999;116:1246–9.
[23] Metchnikoff E. The prolongation of life; Optimistic studies. In: Mitchell, editor. London: William Heinemann; 1907. p. 161–83.
[24] Tissier H. Recherches sur la flore intestinale des nourissons (etat normal et pathologique). Paris: University of Paris; 1906.
[25] Lilly DM, Stillwell RH. Probiotics: growth-promoting factors produced by microorganisms. Science 1965;147:747–8.
[26] Guarner F, Schaafsma G. Probiotics. Int J Food Microbiology 1998;39:237–8.
[27] Karimi O, Pena AS. Probiotics: isolated bacteria strain or mixtures of different strains? Two different approaches in the use of probiotics as therapeutics. Drugs Today (Barc) 2003;39:565–97.
[28] Salminen S, von Wright A, Morelli L, et al. Demonstration of safety of probiotics–a review. Int J Food Microbiol 1998;44:93–106.
[29] Shanahan F. Probiotics and inflammatory bowel disease: is there a scientific rationale? Inflamm Bowel Dis 2000;6:107–15.
[30] Shibolet O, Karmeli F, Eliakim R, et al. Variable response to probiotics in two models of experimental colitis in rats. Inflamm Bowel Des 2002;8:399–408.
[31] Madsen K, Cornish A, Soper P, et al. Probiotic bacteria enhance murine and human intestinal epithelial barrier function. Gastroenterology 2001;121:580–91.

[32] Gionchetti P, Rizzello F, Helwig U, et al. Prophylaxis of pouchitis onset with probiotic therapy: a double-blind, placebo-controlled trial. Gastroenterology 2003;124:1202–9.
[33] Lammers KM, Vergopoulos A, Babel N, et al. Probiotic therapy in the prevention of pouchitis onset: decreased, interleukin-1beta, interleukin-8, and interferon-gamma expression. Inflamm Bowel Dis 2005;11:447–54.
[34] Ulisse S, Gionchetti P, D'Alo S, et al. Expression of cytokines, inducible nitric oxide synthase, and matrix metalloproteinases in pouchitis: effects of probiotic treatment. Am J Gastroenterol 2001;96:2691–9.
[35] Lammers KM, Helwig U, Swennen E, et al. Effect of probiotic strains on interleukin 8 production by HT29/19A cells. Am J Gastroenterol 2002;97:1182–6.
[36] Otte JM, Podolsky DK. Functional modulation of enterocytes by gram-positive and gram-negative microorganisms. Am J Physiol Gastrointest Liver Physiol 2004;286(4):G613–26.
[37] Petrof EO, Kojima K, Ropeleski MJ, et al. Probiotics inhibit nuclear factor-kappaB and induce heat shock proteins in colonic epithelial cells through proteasome inhibition. Gastroenterology 2004;127:1474–87.
[38] Klinman DM, Yi AK, Beaucage SL, et al. CpG motifs present in bacteria DNA rapidly induce lymphocytes to secrete interleukin 6, interleukin 12, and interferon gamma. Proc Natl Acad Sci U S A 1996;93:2879–83.
[39] Krieg AM, Yi AK, Matson S, et al. CpG motifs in bacterial DNA trigger direct B-cell activation. Nature 1995;374:546–9.
[40] Rachmilewitz D, Katakura K, Karmeli F, et al. Toll-like receptor 9 signaling mediates the anti-inflammatory effects of probiotics in murine experimental colitis. Gastroenterology 2004;126:520–8.
[41] Hemmi H, Takeuchi O, Kawai T, et al. A Toll-like receptor recognizes bacterial DNA. Nature 2000;408:740–5.
[42] Jijon H, Backer J, Diaz H, et al. DNA from probiotic bacteria modulates murine and human epithelial and immune function. Gastroenterology 2004;126:1358–73.
[43] Lammers KM, Brigidi P, Vitali B, et al. Immunomodulatory effects of probiotic bacteria DNA: IL-1 and IL-10 response in human peripheral blood mononuclear cells. FEMS Immunol Med Microbiol 2003;38:165–72.
[44] Lammers KM, Hart AL, Brigidi P, et al. DNA derived from probiotic bacteria modulates human dendritic cell phenotype and function. Gastroenterology 2004;(Suppl):A517.
[45] Hart AL, Lammers K, Brigidi P, et al. Modulation of human dendritic cell phenotype and function by probiotic bacteria. Gut 2004;53:1602–9.
[46] Drakes M, Blanchard T, Czinn S. Bacterial probiotic modulation of dendritic cells. Infect Immun 2004;72:3299–309.
[47] Schmitz H, Barmeyer C, Fromm H, et al. Altered tight junction structure contributes to the impaired epithelial barrier function in ulcerative colitis. Gastroenterology 1999;116: 301–9.
[48] Mack DR, Michail S, Wei S, et al. Probiotics inhibit enteropathogenic *E coli* adherence in vitro by inducing intestinal mucin gene expression. Am J Physiol 1999;276:G941–50.
[49] Bernet MF, Brassart D, Neeser JR, et al. Lactobacillus acidophilus LA1 binds to cultured human intestinal cell lines and inhibits cell attachment and cell invasion by enterovirulent bacteria. Gut 1994;35:483–9.
[50] Willemsen LE, Koetsier MA, van Deventer SJ, et al. Short chain fatty acids stimulate epithelial mucin 2 expression through differential effects on prostaglandin E(1) and E(2) production by intestinal myofibroblasts. Gut 2003;52:1442–7.
[51] Venturi A, Gionchetti P, Rizzello F, et al. Impact on the composition of the faecal flora by a new probiotic preparation: preliminary data on maintenance treatment of patients with ulcerative colitis. Aliment Pharmacol Ther 1999;13:1103–8.
[52] Fedorak RN, Gionchetti P, Campieri M, et al. VSL#3 probiotic mixture induces remission in patients with active ulcerative colitis. Gastroenterology 2003;124:A377.

[53] Gionchetti P, Rizzello F, Venturi A, et al. Oral bacteriotherapy as maintenance treatment in patients with chronic pouchitis: a double-blind, placebo-controlled trial. Gastroenterology 2000;119:305–9.
[54] Mimura T, Rizzello F, Helwig U, et al. Once daily high dose probiotic therapy (VSL#3) for maintaining remission in recurrent or refractory pouchitis. Gut 2004;53:108–14.
[55] Gionchetti P, Amadini C, Rizzello F, et al. Probiotics for the treatment of postoperative complications following intestinal surgery. Best Pract Res Clin Gastroenterol 2003; 17:821–31.

Gastroenterol Clin N Am 34 (2005) 515–532

GASTROENTEROLOGY CLINICS
OF NORTH AMERICA

Saccharomyces boulardii: Basic Science and Clinical Applications in Gastroenterology

Jean-Paul Buts, MD, PhD[a,*], Paul Bernasconi, MD[b]

[a]Department of Pediatrics, Cliniques Universitaires St-Luc, Catholic University of Louvain, 10, Avenue Hippocrate, B-1200 Brussels, Belgium
[b]Biocodex, 7, Avenue Gallieni, 94250 Gentilly, France

Saccharomyces boulardii is a nonpathogenic yeast widely prescribed in a lyophilized form in many countries of the world and used in adults and children as a biotherapeutic agent [1,2]. Controlled clinical trials have demonstrated the efficacy of *S boulardii* for preventing or treating several intestinal disorders including antibiotic-associated diarrhea [3,4], recurrent *Clostridium difficile* disease [5,6], *C difficile*-associated enterocolopathies in infants and children [7], acute diarrhea in children [8] and in adults [9], traveler's diarrhea [10,11], diarrhea in tube-fed patients [12], AIDS-related diarrhea [13], and relapses of Crohn's disease [14] or ulcerative colitis [15].

S boulardii differs from *S cerevisiae* by several taxonomic, metabolic, and genetic properties [16,17]. Lyophilized *S boulardii* is obtained by cryodessication, a technique that allows it to preserve viability and stability. The pharmacodynamic properties of *S boulardii* are linked mostly to the revitalization capacity of the yeast after oral administration to the host. *S boulardii* is genetically resistant to all antibacterial antibiotics [18,19], gastric acidity [18,19], and proteolysis. After oral administration of lyophilized *S boulardii*, steady-state concentrations of viable yeast cells are achieved within a mean of 3 days, and the cells are cleared from the stools within 2 to 5 days after discontinuation. Thus, *S boulardii* reaches rapidly in the gastrointestinal (GI) tract high concentrations and remains in a viable (Fig. 1) form throughout the bowel [20]. *S boulardii* exerts several beneficial effects on the host GI tract including protective effects against enteric pathogens such as *Vibrio cholerae* [21], *C difficile* [22,23] and *Escherichia coli* [24,25], lysis of enterotoxins and their binding to intestinal receptors (*C difficile*) [22], stimulation of immune host

*Corresponding author. *E-mail address*: buts@gype.ucl.ac.be (J.-P. Buts).

0889-8553/05/$ – see front matter
doi:10.1016/j.gtc.2005.05.009

gastro.theclinics.com

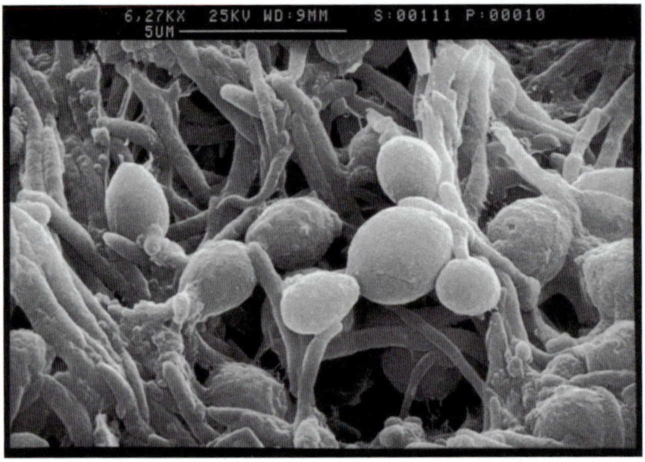

Fig. 1. Scanning electron micrograph of the cecum of mice treated with *Saccharomyces boulardii*.

defenses [26], inhibition of the inflammatory response induced by enterotoxins [25], and enhancement of trophic factors such as brush border membrane enzymes and nutrient transporters [27,28]. A summary of these beneficial effects is shown in Table 1.

According to these pharmacological and clinical data, *S boulardii* is a biotherapeutic medication clearly distinct from nutrient probiotic foods containing various species of microorganisms or of substances able to modify the microflora and given in fermented milks or in yogurts to improve host physiology [11].

This article focuses on the clinical applications of the administration of lyophilized *S boulardii* and explains its beneficial effects on the basis of basic experimental studies.

Table 1
Mechanisms of action of *Saccharomyces boulardii*

Mechanisms	Answers
Production of antimicrobial substances	Not known
Interaction with host microflora	Yes
Inactivation of bacterial toxins	Yes
Inhibition of toxins binding to their intestinal receptors	Yes
Antisecretory effects (Na^+, Cl^-)	Yes
Stimulation of host immune system	Yes
Inhibition of the inflammatory response induced by enterotoxins	Yes
Trophic effects on intestinal mucosa	Yes

EFFECTS OF SACCHAROMYCES BOULARDII ON EXPERIMENTAL INTESTINAL INFECTIONS

Effects have been assessed on different models of intestinal infections or colonization by different pathogenic microorganisms. They have been reviewed recently by Czerucka and Rampal [29].

Clostridium difficile

Administration of *S boulardii* significantly reduces mortality caused by *C difficile* colitis in hamsters treated with clindamycin [30] and in mice inoculated orally with *C difficile* [31] or directly with toxin A and B of this pathogen [32].

The absence of intestinal or colonic lesions in animals protected by *S boulardii* has been observed using optical [33] and electron [34] microscopy and has been confirmed on an in vitro model of a cell line (IRD 98) belonging to rat epithelial intestinal cells [35]. The protective effect against *C difficile* enterotoxigenicity is more marked when *S boulardii* is given preventively and continuously [30,31,36]. This protective effect disappears when yeast cells are killed by amphotericin-B or by heating and is proportional to the oral dose administered and to the viability of *S boulardii* in the GI tract [36]. Elmer and Corthier have documented a significant correlation between the survival rates of mice infected with *C difficile* and the oral dose of *S boulardii* given [36]. If the dose of *S boulardii* was increased from 3×10^8 to 3.3×10^{10} cells/mL drinking water, survival rates increased linearly from 0% to 85% [36].

In addition, the protective effect of *S boulardii* was correlated with the intestinal concentration of *C difficile* and the levels of toxin A or toxin B [37].

The protective effect of *S boulardii* against intestinal infections due to *C difficile* results from several complementary mechanisms.

The first of these mechanisms is the secretion in vivo by the yeast of a 54 kDa protease, which inhibits the enterotoxinic and cytotoxic effects of toxins A and B. The intestinal receptor for toxin A of *C difficile* is a glycoprotein of high molecular weight, sensitive to proteases. Pothoulakis and colleagues [22,37] have demonstrated that *S boulardii* produces in vivo an original protease of 54 kDa, able to decrease intestinal hydroelectrolytic secretions in an isolated loop of rabbit but being without effect on the lesions caused by *C difficile* on other cell lines than intestinal epithelial cells, such as lung fibroblasts (SMR-90) or basophilic leukemia cells of rats [RBL]. This protease of *S boulardii* inhibits binding of purified [H^3]-enterotoxin A to brush border membranes by 37%, reduces by 55% hydroelectrolytic secretions induced by enterotoxin A, and decreases by 93% intestinal permeability to mannitol. In addition, after oral administration of *S boulardii* to rats, during 3 days oral administration, enterotoxin A had no effect on hydroelectrolytic secretions and on intestinal permeability. Castagliulo and colleagues [23,37] have confirmed that the partially purified 54 kDa protease could lyse toxin A promptly and specifically and partially destroy the intestinal receptor, resulting in an inhibition of binding of toxin A and B on their intestinal receptors.

The second mechanism is the stimulation of the immune response against toxin A. In mice inoculated orally with a toxoïd of toxin A, administration of *S boulardii* results in an important stimulation of the specific immune response of secretory IgA and IgM antitoxin A in the serum [38].

The third mechanism is inhibition of the adherence of *C difficile* to intestinal cells by *S boulardii*. This has been documented in vitro on a cell line [39].

Vibrio cholerae

In a model of the isolated jejunal loop of rats, administration of *S boulardii* reduced hydrosaline hypersecretion induced by cholera toxin significantly [40]. This inhibitory effect has been confirmed in vitro on several intestinal epithelial cell lines of rats IRD 98 [35], IEC 17 [35] and IEC6 [21]. This effect of *S boulardii* is associated with a 50% decrease in the activity of cyclic adenosine monophosphate (AMP) induced by the cholera toxin. The protective effect disappears after destruction of the yeast by heating. The supernatant of the culture media of *S boulardii* (also called conditioned medium: CM) is able by itself to decrease, c-AMP activities induced by cholera toxin in IEC 6 cells in a dose-dependent fashion. This effect disappears when the medium is denatured by heating, hydrolyzed by trypsin, or treated with trichloroacetic acid, suggesting the presence of a protein secreted by the yeast in the medium. This protease has been identified on polyacrylamide gels as a 120 kDa protein [21]. This protein and the CM, however, do not change the binding parameters of iodinated cholera toxin on the intestinal receptor [41] and do not show a proteolytic activity on the cholera toxin [21]. These observations have prompted Czerucka and colleagues [42] to study the effects of CM of *S boulardii* on the intracellular signal pathway regulating the secretion of chloride.

This effect of CM on the transport of chloride has been studied on T84 cells derived from a human carcinoma. These cells present the characteristics of crypt cells secreting chloride by the Na^+ pump, K^+ ATPase, the co-transporter $Na^+/K^+/Cl^-$ and the K^+ channel located at the basolateral side of the cell and the chloride channel located at the apical side. All of these exchange mechanisms play a role in the secretion of chloride induced by agonists acting by means of the c-AMP such as forskolin, vasoactive intestinal polypeptide (VIP), and prostaglandins (PGE), or by means of the Ca^{2+} pathway such as carbachol [43]. Chloride secretion induced 60 minutes after exposure to cholera toxin has been quantified by the technique of Venglarik and colleagues [44].

In the presence of *S boulardii* CM, chloride secretion and the levels of c-AMP stimulated by cholera toxin in T84 cells were reduced to the control level. Furthermore, CM of *S boulardii* decreased chloride secretion enhanced by VIP and PGE2 only if cells had been exposed during 60 minutes to CM preparation of *S boulardii* before the incubation with the agonists. One hour incubation with CM of *S boulardii* was also necessary to decrease chloride secretion induced by carbachol without decrease in the intracellular concentration of inositol-triphosphate. CM of *S boulardii* seems to generate in T84 cells a mediator able to dissociate ionic channels and intracellular calcium levels. In summary, these data suggest that *S boulardii* secretes a 120-kDa protein that acts directly on enterocytes and regulates in the cell the pathways controlling chloride secretion.

Escherichia coli

On the model of acute diarrhea induced in mice by the thermostable enterotoxin of enterotoxigenic *Escherichia coli* (ETEC), the administration of a suspension of *S boulardii* exerts a significant inhibitory effect on intestinal hydroelectrolytic secretion [45].

Enteropathogenic *E coli* (EPEC) are bacterial pathogens predominating in infantile gastroenteritis, especially in developing countries. Enterohemorrhagic *E coli* (EHEC) are implicated in alimentary contaminations occurring in industrialized countries (United States, Canada, Japan) and produce hemorrhagic colitis associated in some cases with the hemolytic-uremic syndrome [45]. These pathogens secrete a shiga-like toxin that inhibits protein synthesis in intestinal cells but does not seem to be involved in the intestinal pathology caused by EHEC. The pathogenic sequence of EPEC and EHEC begins with their adhesion to the intestinal mucosa resulting in patchy destructions of microvillus membranes, accumulation of actin fibers under the site of bacterial attachment [46], and, as a result, disruption of the tight junctions leading to an increase in intestinal permeability with disappearance of the intestinal barrier [46].

These changes are associated with a translocation and migration of polynuclear cells and with the synthesis of proinflammatory cytokines (interleukin [IL]-8) [47]. Some responses of the enterocytes to infection have been correlated with the activation of the phosphorylation of intracellular signal proteins. For instance, phosphorylation of the myosin light chain (MLC) has been implicated directly in the disruption of tight junctions during acute infections by EPEC [48].

Oral administration of *S boulardii* during infection of T84 cells with EPEC producing a shiga-like toxin (strain EDL 931) reactivates the barrier function as measured by the transepithelial resistance, the absorption of inulin, or the immunolocalization of protein ZO-1.

In addition, cleavage of caspase 32 shows that *S boulardii* is able to delay the apoptosis processes resulting from the infection of T84 cells by EPEC pathogens. The yeast has no effect on the amount of EPEC or EHEC bacteria adherent to enterocytes. By contrast, *S boulardii* inhibits the phosphorylation of several signaling molecules, normally phosphorylated in the response of an infection [24,49].

Infection of T84 cells by EPEC or EHEC pathogens induces the activation of the three pathways of MAP (mitogen activating protein) kinases: the extracellular-regulating protein kinase 1 and 2 (ERK-1 and ERK-2), p38 MAP kinase, and c-JUN-kinase. The activation by phosphorylation of ERK-1,2 is associated directly with internalization of the pathogens. The presence of *S boulardii* significantly decreases the number of bacteria internalized and modulates the phosphorylation levels of ERKs-1,2 [25,50] on serine–threonine residues. Under normal physiological conditions, activation of ERK-1 and -2 by diphosphorylation is the result of a growth factor such as the action of insulin or IgF-1 [51,52]. Diphosphorylation of p38 MAP kinase is involved in the process of apoptosis [51]. During the infection of invasive pathogens such as *Salmonella typhimurium* [53] and *Listeria monocytogenes* [54], synthesis and activation of MAP-kinases have been incriminated in the invasion process and in the synthesis of inflammatory cytokines. Czerucka and colleagues [25] have demonstrated that the MAP-kinases pathways stimulated by EPEC and EHEC infection and the activation of the transcription factor NF-kB induced by EHEC are involved in the synthesis of IL-8. *S boulardii* inhibits these pathways and the synthesis of IL-8 during intestinal infection by these pathogens [25,50]. The yeast also inhibits apoptosis induced by EHEC and reduces the production of TNF-α.

Other enteropathogenic bacteria

Administration of a single dose of 10 mg *S boulardii* to gnotobiotic mice first inoculated with a suspension of *Shigella flexneri* or *Salmonella typhimurium* produced a protective effect against mortality (*Shigella flexneri*) and severity of intestinal lesions (*Salmonella typhimurium*) induced by these enteropathogens [53–55].

EFFECTS OF SACCHAROMYCES BOULARDII ON THE GASTROINTESTINAL TRACT

These effects have been reviewed recently [11,19].

Trophic effects

Oral administration of *S boulardii* to human volunteers during 8 days had no effect on intestinal morphology, villus height, crypt depth, or cellularity of the lamina propria [27]. Likewise, electron microscopy of intestinal epithelial cells of rats treated with *S boulardii* showed no intracellular translocation of yeasts and no morphological change of microvilli, villus, and crypt structures [27]. Using a three-dimensional technique of microdensification on human intestinal biopsies, Jahn and colleagues [56] confirmed the authors' findings that after an oral treatment with *S boulardii*, there was no change in villus surface, microvillus structure, and crypt depth. In a study conducted in human volunteers and in growing rats, Buts and colleagues [27] showed that compared with the initial biopsies taken at day 0, the biopsies of human volunteers treated with *S boulardii* presented significant increases in the specific and total activity of sucrase–isomaltase (+82%), lactase (+77%), and maltase–glucoamylase (+75%) after 8 days of oral treatment. Likewise, in 30-day-old growing rats treated during 14 days with *S boulardii*, the specific and total activities of these enzymes were significantly increased compared with a control group treated with a placebo. In agreement with the authors' findings, Jahne and colleagues [56], using an in situ technique to measure microvillus enzyme activities on histological specimens from frozen intestinal biopsies, reported that after treatment with *S boulardii*, lactase, α-glucosidases, and alkaline phosphatase activities were increased by +22% to +55% compared with the basal enzyme activities measured before treatment. From these findings, it appears that *S boulardii* enhances, in human volunteers and in rats, intestinal enzyme activities that assume nutrient degradation and absorption that are frequently altered in acute or chronic enteropathies. *S boulardii* secretes in the endoluminal compartment a sucrase at levels of activities so high that *S boulardii* may be an efficient treatment of congenital sucrase–isomaltase deficiency [57]. A recent study from the authors' laboratory [58] demonstrated that oral administration of *S boulardii* also produces a leucine aminopeptidase that belongs to the family of zinc-metalloproteases and that enhances proteolysis of small N-terminal peptides in the intestinal lumen. This enzyme could help reduce the occurrence of allergenization to dietary proteins, especially during the healing phase of acute gastroenteritis, thereby reducing the likelihood of developing chronic persistent or protracted diarrhea after an acute intestinal infection.

In another recent study [28], the authors showed that oral administration of *S boulardii* to rats during 8 days after a proximal enterectomy of 60% not only enhanced the activity of disaccharidases but also enhanced the absorption of D-glucose coupled to Na^+ by the symport glucose/Na^+.

The absorption of D-glucose measured in vitro on preparations of brush border vesicles was enhanced markedly in relation to the incubation time and glucose concentration in the incubation media, compared with resected rats placebo-treated and with rats that underwent a single transection. In addition, the expression of sodium-glucose cotransporter-1 (SGLT-1) in the brush border membrane, a protein of 70-kDa, when measured by autoradiographic screening, was much more marked in rats resected and treated with *S boulardii* than in the two other groups. These data underline the interest of *S boulardii* for acute gastroenteritis when SGLT-1 is used to reabsorb water and electrolytes with salt replacement therapy to prevent or to treat moderate dehydration. Similar data have been published by Zaouche and colleagues [59].

In addition to the stimulation of brush border enzymes and transporters, *S boulardii* enhances expression of the receptor for polymeric immunoglobulins [26] and the secretion of secretory IgA [26] (s-IgA), whose intraluminal function and intracellular synthesis differ quite totally. Because yeast cells do not penetrate into enterocytes [27] the authors have assessed the potential influence of trophic factors secreted by *S boulardii* cells during their intestinal transit.

Table 2 summarizes the intestinal functions stimulated by oral administration of *S boulardii* and the enzymatic and trophic factors secreted by the yeast during its transit identified up to now.

As shown in Table 3, determinations by high-performance liquid chromatography (HPLC) of polyamines on lyophilized preparations of yeast cells revealed substantial amounts of these substances, totaling 679 nmol/100 mg of lyophilized *S boulardii*, mainly spermidine (55%) and spermine (43%), with very low concentrations of putrescine (1.4%) [60]. Theoretically, these amounts of polyamines can influence the intestinal expression of brush border glycoproteins such as hydrolases, proteases, and transport carriers. Indeed,

Table 2
Trophic effects of *Saccharomyces boulardii*

Factor	Stimulated	Secreted
Sucrase	+++	+++
Maltase-glucoamylase	+++	No
Lactase	++	No
Aminopeptidase	++	++
Alkaline phosphatase	++	+
Protein phosphatase	No	++
Trehalose	?	?
Carboxypeptidase	?	?
Angiotensinase	?	?
DPP IV	?	?
SGLT-1	+++	No
Receptor of polymeric immunoglobulins	+++	No
s-IgA	+++	No

Abbreviations: DPP IV, dipeptidyl peptidase IV; SGLT-1, sodium-glucose cotransporter-one; s-IgA, secretory IgA.
+++ markedly.
++ moderately.
+ slightly.

a marked stimulation of disaccharidase activity, aminopeptidase and the endoluminal secretion of secretory IgA has been observed in the small intestine of infant rats before weaning in response to oral ingestion of spermine and spermidine equal to 1000 nmol of purified polyamines per day [60]. When rats were treated with an amount of spermine (500 nmol/d) equivalent to the content of polyamines of the yeast lyophilized preparation (679 nmol/100 g), similar enzymatic responses were observed, including significant increases in the specific and total activities of sucrase (2.5-fold increase) and of maltase (+24%). In response to 1000 nmol spermine, the stimulation of the enzyme activities was proportionally greater with increases in the activity of sucrase (4.6-fold increase) and of maltase (+70%). Likewise, weaned rats treated with *S boulardii* or with equivalent amounts of spermine (500 nmol) presented parallel

Table 3
Polyamine concentrations in lyophilised preparation of *Saccharomyces boulardii*

	S boulardii nanomoles/mg n = 8	*S boulardii* nmol/mg protein n = 8
Putrescine	0.095 ± 0.014	0.28 ± 0.05
Spermidine	3.766 ± 0.328	10.9 ± 0.84
Spermine	2.930 ± 0.268	8.42 ± 0.67
Total	6.790	19.60

Values are means ± SD [77].

and significant increases in the specific activity of sucrase (+157%) and of maltase (+47.5%) [60].

Thus, oral administration of 100 mg lyophilized *S boulardii* containing 679 nmol of polyamines compared with 500 nmol of purified spermine given to suckling rats or to weaned rats results in similar enzymatic responses. The stimulation of sucrase and maltase by oral administration of spermine is a dose-dependent effect that for sucrase and maltase is more sensitive than for lactase, aminopeptidase, and alkaline phosphatase, and which becomes detectable for doses of spermine exceeding 250 nmol/d. In addition, both *S boulardii* [60] and purified spermine given at 500 nmol per day can stimulate significantly the intestinal production of the intestinal receptor for immunoglobulins in weanling rats between postpartum days 20 and 40.

Besides changes in enzyme activities, oral treatment with *S boulardii* produced parallel changes in the concentrations of polyamines in intestinal mucosa (+21.4%) and in endoluminal fluid (+48% to +316% in the jejunum and +60.8% to +150% in the ileum). Variations in concentration of the three polyamines measured by HPLC in the intestinal mucosa were proportional to their respective concentrations measured in the lyophilized preparation of the yeast. Indeed, spermine and spermidine, which represented respectively 44% and 55% of the total of polyamines produced by *S boulardii*, were increased in similar proportions in the intestinal mucosa (+21.4% and +21.9%).

In concordance with the very low amounts of putrescine produced by the yeast (1.4%), putrescine concentrations in the small bowel mucosa varied very little and were not affected by the oral treatment.

Experimental data indicate that the transepithelial transport of polyamines by brush border membrane vesicles is a selective and saturable process, largely dependent upon intraluminal concentrations [60].

In samples of jejunal and ileal endoluminal fluid rinsed and filtered to discard yeast cells, spermidine and spermine levels were increased after oral treatment by +48% to +316% in treated rats compared with controls, while the levels of putrescine were not enhanced significantly.

In summary, the data show that at the doses used, the lyophilized preparation of *S boulardii* exerts trophic effects on the small intestinal mucosa that likely are mediated, at least in part, by the endoluminal production of spermine and spermidine. These polycationic substances seem to be released by yeast cells during their catabolic phase rather than being secreted by viable yeast cells during their endoluminal transit. Indeed, only traces of putrescine have been detected in the culture media of yeast after 96 hours of exponential growth without evidence of spermine or of spermidine in the media. Because the physiological effects of exogenous polyamines on intestinal cell maturation, intestinal enzyme expression, transport mechanisms, crypt cell division, DNA synthesis and gene transcription are established, the substantial amount of polyamines delivered by *S boulardii* could have important clinical implications. Although the use of *S boulardii* as a biotherapeutic agent has been documented in acute gastroenteritis, in antibiotic-associated diarrhea, and in enterocolopathies caused by *C difficile*, the potential trophic effects of the yeast for preventing chronic persistent diarrhea, chronic protracted diarrhea, allergenization to dietary proteins, and for the substitutive treatment of congenital enzymatic deficiencies (sucrase–isomaltase, trehalase) warrant further investigations.

Beside the trophic effects mediated by polyamines, an important matter of study has been raised by the effects of *S boulardii* on the metabolism of short-chain fatty acids (SCFAs), especially butyrate, acetate, and propionate. SCFAs are among the most important metabolites produced by the anaerobic colonic microflora, and they play a major role in colonic reabsorption of water and electrolytes [61]. In a porcine model of bacterial ecosystem, administration of *S boulardii* was able to restore the production of SCFA, at normal levels, which were depressed by the decrease of bacterial mass due to clindamycin treatment [62].

This property can play a role in acute gastroenteritis and explain in part the antidiarrheal effect of *S boulardii*, because SCFA are decreased in relation to microflora alterations in antibiotic-associated diarrhea and diarrhea occurring during continuous enteral feeding.

Antisecretory effects

S boulardii reduces hydroelectrolytic secretions (−39%) and permeability of the mucosa to mannitol (−65%) induced by toxin A of *C difficile* in the rat ileum [22,23].

Furthermore, *S boulardii* inhibits the secretion of chloride induced by the signal transduction pathway of cyclic AMP and the calcium pathway in epithelial villus cells [42,63].

Lastly, when administered preventively in rats, *S boulardii* exerts a marked effect on the secretory diarrhea induced by castor oil, the effect being dose-dependent [64]. This effect is inhibited by L-arginine, suggesting a role of the nitric oxide (NO) pathway in this mechanism.

Immunological effects

In growing rats, *S boulardii* produces a marked increase (+57%) in the concentrations of secretory s-IgA in the intestinal lumen and the concentration of the receptor for polymeric immunoglobulins in crypt cells (+63%) [26]. A similar effect has been documented in BALB/c mice orally treated by a toxoïd of *C difficile* toxin A, resulting in a strong increase in total s-IgA production (1.8-fold) and in the levels of sectretory s-IgA specifically directed against toxin A (4.4-fold) [38].

Anti-inflammatory effects

Studies on the protective mechanisms of *S boulardii* on infections EPEC and EHEC have shown that *S boulardii* inhibited the pathways of MAP kinases and of NF-kB, leading to apoptosis and secretion of proinflammatory cytokine IL-8 [49]. Likewise, Sougioultzis and colleagues [65] have shown the production by the yeast of a hydrosoluble factor that is able to inhibit NF-kB when intestinal cells are exposed to the toxin A of *C difficile*. These findings emphasize the importance of developing clinical studies of *S boulardii* for treating inflammatory bowel diseases (IBDs).

In summary, the mechanisms of the antidiarrheal effects of *S boulardii* may be attributed to two main pharmacodynamic properties of the yeast cells: (1) inhibition of some bacterial toxins or of their pathogenic effects and (2) direct effects on intestinal mucosa, including trophic effects, antisecretory effects, a stimulation of intestinal immunity, and an anti-inflammatory effect.

The clinical efficacy of lyophilized *S boulardii* has been assessed in several intestinal diseases.

ANTIBIOTIC-ASSOCIATED DIARRHEA AND *CLOSTRIDIUM DIFFICILE* ENTEROCOLOPATHIES

Antibiotic-associated diarrhea (AAD) occurs between 5% and 30% of the time during antibiotic treatment. Symptoms range from simple diarrhea without dehydration or hydroelectrolytic disturbances to severe secretory diarrhea that is potentially lethal, with in extreme cases pseudomembranous colitis or enterocolitis. *C difficile* is thought to be responsible for 10% to 30% of AAD and for 95% of pseudomembranous colitis [66]. *C difficile* is also the cause of protracted chronic enterocolopathies in infants [7]. Beside their potentially lethal outcome, AADs caused by *C difficile* overgrowth are characterized by a high frequency of clinical relapses: 20% of relapses after the initial affection and 60% after the first relapse. In some cases, another pathogen, especially *Klebsiella oxytoca* [67], has been incriminated. The risk factors to develop AAD now are understood better. They are linked either to the type and duration of antibiotherapy (long treatments, multiple antibiotics, antibiotics excreted in the bile, antibiotics acting on anaerobes, broad-spectrum antibiotics) or to factors linked to the host (eg, very young or very old persons, surgery, chronic digestive diseases, severity of the causal infection, immunosuppression) [68]. The existence of one or of a combination of these factors of risk may justify prophylactic treatment.

Prevention of antibiotic-associated diarrhea

Three prospective, randomized, placebo-controlled clinical studies have confirmed the efficacy of *S boulardii* for preventing AAD.

In a first study, conducted in 180 hospitalized patients [3], *S boulardii* was given at the dose of 1 g/d during the period of antibiotherapy up to 2 weeks after discontinuation of antibiotics. The occurrence of AAD was 21.8% in the placebo group versus 9.5% in the treated group ($P = .038$).

In a second two-center study of 193 hospitalized patients receiving a β-lactam in combination with another antibiotic or not [4], *S boulardii* was administered at the dose of 1 g/d and was continued during 3 days after discontinuation of antibiotics. Occurrence of AAD was 14.6% in the placebo group versus 7.2% in the group treated with *S boulardii* ($P = .03$).

These data confirm those recorded in a preliminary study conducted in 388 ambulatory patients [69]. Treatment with *S boulardii* at the dose of

200 mg/d during the antibiotherapy (β-lactams or cyclines) decreased the percentage of AAD from 17.5% to 4.5% [69].

Treatment of recurrent diarrhea and colitis due to Clostridium difficile

Several clinical trials aiming to assess the efficacy of *S boulardii* for treating relapses of colitis and diarrhea caused by *C difficile* have been performed.

In a preliminary open study, conducted in 13 patients presenting multiple relapses of pseudomembranous colitis caused by *C difficile* [70], administration of *S boulardii* at 1 g/d during 1 month in association with vancomycin prevented further relapses in 11 patients.

A multi-center prospective, randomized, placebo-controlled study in 124 patients treated with vancomycin or metronidazole for an episode of colitis or diarrhea caused by *C difficile* [5] has assessed the effect of *S boulardii* on the occurrence of further relapses. In patients having had at least one relapse ($n = 60$), administration of *S boulardii* reduced the occurrence of further relapses significantly: 64.7% relapses in the placebo group versus 34.6% relapses in the group treated with *S boulardii* ($P = .04$).

Another multi-center, prospective and placebo-controlled study has been conducted in 170 patients presenting a chronic intestinal infection caused by multi-resistant *C difficile* strains. That study has shown that the treatment of choice was the combination of high doses of vancomycin with *S boulardii* [6].

Treatment of enterocolopathies caused by *Clostridium difficile* in infants

In infants and children, intestinal overgrowth of a pathogenic strain of *C difficile* rarely results in pseudomembranous colitis. More often, it produces an acute enteritis or a chronic enterocolopathy. In an open trial of 19 infants and children presenting with chronic diarrhea or chronic protracted enterocolopathy caused by toxinogenic resistant *C difficile*, *S boulardii* treatment at the dose of 500 to 1000 mg/d during 15 days resulted in a prompt disappearance of clinical symptoms in 18 patients, with negativation of toxin B in 16 patients [7]. Most of the patients had been treated before with one or several antibiotics.

ACUTE GASTROENTERITIS

A prospective, randomized, placebo-controlled study evaluated *S boulardii* at 500 mg/d in acute gastroenteritis of moderate intensity occurring in young children [71]. At day 1 and day 4 in the treated ($n = 19$) and in the control group ($n = 19$), there was a significant difference in four criteria: number, weight, consistency of stools, and transit time measured by the red carmine method [71].

In another placebo-controlled study of 130 children aged from 3 months to 3 years, administration of *S boulardii* at 300 mg/d significantly reduced the frequency of stools after 48 hours of treatment. At 48 and 96 hours of treatment, the amount of clinical healing was significantly higher in the treated group [9] than in the placebo group.

In adults, a prospective placebo-controlled study was performed in 92 patients with acute gastroenteritis [8]. Administration of *S boulardii* reduced symptoms and diarrhea significantly after 48 hours ($P = .035$), measured by a score integrating symptoms, number, and consistency of stools.

TRAVELER'S DIARRHEA

This is the most frequent disorder occurring in people traveling in tropical or subtropical countries.

The efficacy of *S boulardii* for preventing traveler's diarrhea has been demonstrated in a large cohort of travelers going to different continents. A total of 1016 persons have been questioned to evaluate the efficacy of *S boulardii*. Treatment with yeast cells has been assumed 5 days before the travel and during all the trip.

Patients were assigned to one of three groups. The first group received a placebo; the second group received *S boulardii* at 250 mg/d, and the third group received *S boulardii* at 1000 mg/d.

The incidence of diarrhea was 39.1%, 34.4% ($P = .019$ versus placebo), and 28.7%, respectively ($P = .005$ versus placebo) [10].

DIARRHEA ASSOCIATED WITH CONTINUOUS ENTERAL FEEDING

Three randomized, prospective and placebo-controlled studies have documented an efficacy of *S boulardii* for preventing diarrhea that occurs during continuous enteral feeding.

In patients hospitalized in a high care medical unit, administration of *S boulardii* at 500 mg/l of nutritive solution reduced the incidence of diarrheal days from 16.9% to 8.7% ($P < .001$) [72].

In another study of patients hospitalized in a sterile unit burn, preventive treatment with *S boulardii* at the dose of 2 g/d reduced the incidence of diarrheal days from 9.1% to 1.5% ($P < .001$), with a significant improvement in digestive tolerance to the enteral diet and a significant increase in caloric intake [73].

Lastly, a multi-center study (11 centers) conducted in 128 patients has confirmed the efficacy of *S boulardii* given at 2 g/d for preventing diarrhea in patients hospitalized in a intensive care unit and nourished with enteral feeding [12]. The authors observed a significant reduction of the number of days with diarrhea (14% versus 19% in the placebo group, $P < .01$) in the group treated with *S boulardii* (relative efficacy of 29%).

DIARRHEA IN PATIENTS WITH AIDS

In an open study on 17 patients presenting chronic diarrhea associated with AIDS, *S boulardii* given at 3 g/d during 15 days exerted an antidiarrheal effect, with diminution of the mean stool output from 9.0 to 2.1 per day [13]. A double-blind, placebo-controlled study conducted in 35 patients has confirmed these results. After 1 week of treatment, diarrhea was controlled in 61% of treated patients compared with 12% in the placebo group (<.002). The number, weight, and volume of stools; abdominal pain; body weight; and the Karnofsky quality of life index also were improved significantly in patients treated with *S boulardii* [74].

IRRITABLE BOWEL SYNDROME

A prospective placebo-controlled, clinical trial composed of 34 patients with irritable bowel syndrome with predominant episodes of diarrhea has shown that treatment with *S boulardii* has a significant effect on the number and consistency of stools ($P < .05$) after 1 month of treatment [75].

CHRONIC INFLAMMATORY BOWEL DISEASES

A prospective randomized clinical trial has assessed the efficacy of *S boulardii* (1 g/d) with mesalamine (2 g/d) versus mesalamine alone (3 g/d) for preventing relapses of Crohn's disease [14] in adults in remission for the disease. After 6 months, the incidence of relapses reached 37.5% in the group receiving mesalamine alone (n = 16) versus 6.25% in the group treated with mesalamine and *S boulardii* (n = 16, $P = .04$).

A recent pilot study showed interesting data for treating moderate forms of ulcerative colitis [15].

SAFETY OF USE

Lyophilized *S boulardii* is tolerated well under usual conditions of use. In very rare cases, *S boulardii* fungemias have been reported. In documented cases, all patients had an indwelling central venous catheter [76].

S boulardii fungemia disappeared, either spontaneously with discontinuation of the product or as a result of antifungal treatment. In some cases, the removal of the central venous catheter was necessary. Colonization of the central venous catheter during handling of packets or capsules appears to be the most likely involved mechanism. Consequently, administration of lyophilized *S boulardii* is contraindicated in patients with an indwelling central venous catheter [77].

SUMMARY

Because of major advances in understanding in intestinal physiology and in the role of intestinal microflora, many of the mechanisms of action of *S boulardii* have been elucidated: secretion in vivo of proteases and substances inhibiting bacterial toxins or their pathogenic effects, trophic effects, antisecretory effects, immunostimulatory effects and anti-inflammatory effects.

The efficacy of lyophilized *S boulardii* has been established by clinical studies, conducted prospectively and placebo-controlled, in several pathological conditions such as AAD, *C difficile* enterocolopathies, pseudomembranous enterocolitis, acute gastroenteritis, traveler's diarrhea, chronic diarrhea of patients with AIDS, and IBDs.

The recent discovery of a clear anti-inflammatory effect of this biotherapeutic agent warrants further scientific studies and clinical trials in IBD.

References

[1] McFarland LV, Bernasconi P. *Saccharomyces boulardii*: a review of an innovative biotherapeutic agent. Microb Ecol Health Dis 1993;6:151–71.
[2] Elmer GW, Surawicz CM, McFarland LV. Biotherapeutic agents. A neglected modality for the treatment and prevention of selected intestinal and vaginal secretions. JAMA 1996;275:870–6.
[3] Surawicz CM, Elmer GW, Speelman P, et al. Prevention of antibiotic associated diarrhea by *Saccharomyces boulardii*: a prospective study. Gastroenterology 1989;96:981–8.
[4] McFarland LV, Surawicz CM, Greenberg RN. Prevention of β-lactam associated diarrhea by *Saccharomyces boulardii* compared with placebo. Am J Gastroenterol 1995; 90:439–48.
[5] McFarland LV, Surawicz CM, Greenberg RN. A randomised placebo-controlled trial of *Saccharomyces boulardii* in combination with standard antibiotics for *Clostridium difficile* disease. JAMA 1994;271:1913–8.
[6] Surawicz CM, McFarland LV, Greenberg R. The search for better treatment for *Clostridium difficile* disease: use of high-dose vancomycin combined with *Saccharomyces boulardii*. Clin Infect Dis 2000;31:1012–7.
[7] Buts JP, Corthier G, Delmée M. *Saccharomyces boulardii* for *Clostridium difficile*-associated enterocolopathies in infants. J Pediatr Gastroenterol Nutr 1993;16:1497–504.
[8] Höchter W, Chase D, Hegenhoff G. *Saccharomyces boulardii* in treatment of acute adult diarrhea. Efficacy and tolerance of treatment. MMW Münch Med Wochenschr 1990; 132:188–92.
[9] Cetina-Sauri G, Sierra Basto G. Evaluation thérapeutique du *Saccharomyces boulardii* chez des enfants souffrant de diarrhée aiguë. Ann Pediatr 1994;41:397–400.
[10] Kollaritsch H, Holst H, Grobara P. Prophylaxe der Reisediarrhoë mith *Saccharomyces boulardii*. Fortschr Med 1993;111:152–6.
[11] Buts JP. Exemple d'un médicament probiotique: *Saccharomyces boulardii* lyophilisé. In: Rambaud JC, Buts JP, Corthier G, et al, editors. Flore microbienne intestinale: physiologie et pathologie digestives. Montrouge, (France): John Libbey Eurotext; 2004. p. 221–44.
[12] Bleichner G, Bléhaut H, Mentec H, Moyse D. *Saccharomyces boulardii* prevents diarrhea in critically ill tube-fed patients. A multi-center, randomised, double-blind, placebo-controlled trial. Intensive Care Med 1997;23:517–23.
[13] Saint-Marc T, Bléhaut H, Musial C, et al. Efficacité de *Saccharomyces boulardii* dans le traitement des diarrhées du SIDA. Annales de Medicine Interne 1991;142:64–5.
[14] Guslandi M, Mezzi G, Sorghi M, Testoni PA. *Saccharomyces boulardii* in maintenance treatment of Crohn's disease. Dig Dis Sci 2000;45:1462–4.
[15] Guslandi M, Giollo P, Testoni PA. A pilot trial of *Saccharomyces boulardii* in ulcerative colitis. Eur J Gastroenterol Hepatol 2003;15:697–8.
[16] Hennequin C, Thierry A, Richard GF. Microsatellite typing as a new tool for identification of *Saccharomyces cerevisiae* strains. J Clin Microbiol 2001;39:551–9.
[17] Malgoire JY, Bertout S, Renaud F, et al. Typing of *Saccharomyces cerevisiae* clinical strains by using microsatellite sequence polymorphism. J Clin Microbiol 2005;43:1133–7.
[18] Bergogne-Berezin E. Impact écologique de l'antibiothérapie. Place des microorganismes de substitution dans le contrôle des diarrhées et colites associées aux antibiotiques. Presse Méd 1995;24:145–56.
[19] Buts JP. Mechanisms of action of biotherapeutic agents. In: Elmer GW, McFarland LV, Surawicz CM, editors. Biotherapeutic agents and infectious diseases. Totowa (NJ): Humana Press; 1999. p. 27–46.
[20] Bléhaut H, Massot J, Elmer GW, et al. Disposition kinetics of *Saccharomyces boulardii* in man and rat. Biopharm Drug Dispos 1989;10:353–64.

[21] Czerucka D, Roux J, Rampal P. *Saccharomyces boulardii* inhibits secretagogue-mediated adenosine 3′, 5′-cyclic monophosphate induction in intestinal cells. Gastroenterology 1994;106:65–72.
[22] Pothoulakis C, Kelly CP, Joshi MA. *Saccharomyces boulardii* inhibits *Clostridium difficile* toxin A and enterotoxicity in rat ileum. Gastroenterology 1993;103:1108–15.
[23] Castagliulo I, Riegler MF, Valenick L, et al. *Saccharomyces boulardii* protease mediates *Clostridium difficile* toxin A and B effects in human colonic mucosa. Infect Immun 1999;61:302–7.
[24] Czerucka D, Dahan S, Mograbi B, et al. *Saccharomyces boulardii* preserves the barrier function and modulates the signal transduction pathway induced in enteropathogenic *Escherichia coli* infected T84 cells. Infect Immun 2000;68:5998–6004.
[25] Czerucka D, Dahan S, Mograbi B, et al. Implication of mitogen-activated protein kinases in T84 cell responses to enteropathogenic *Escherichia coli* infection. Infect Immun 2001;69: 1298–305.
[26] Buts JP, Bernasconi P, Vaerman JP, et al. Stimulation of secretory IgA and secretory component of immunoglobulins in the small intestine of rats treated with *Saccharomyces boulardii*. Dig Dis Sci 1990;35:251–6.
[27] Buts JP, Bernasconi P, Van Craynest MP, et al. Response of human and rat small intestinal mucosa to oral administration of *Saccharomyces boulardii*. Pediatr Res 1986;20:192–6.
[28] Buts JP, De Keyser N, Marandi S. *Saccharomyces boulardii* upgrades cellular adaptation after proximal enterectomy in rats. Gut 1999;45:89–96.
[29] Czerucka D, Rampal P. Experimental effects of *Saccharomyces boulardii* on diarrheal pathogens. Microbes Infect 2002;4:733–9.
[30] Toothaker RD, Elmer GW. Prevention of clindamycin-induced mortality in hamster by *Saccharomyces boulardii*. Antimicrob Agents Chemother 1984;26:552–6.
[31] Corthier G, Dubois F, Ducluzeau R. Prevention of *Clostridium difficile* induced mortality in gnotobiotic mice by *Saccharomyces boulardii*. Can J Microbiol 1986;32:294–6.
[32] Corthier G, Lucas F, Jouvet S, et al. Effect of oral *Saccharomyces boulardii* treatment on activity *Clostridium difficile* toxins in mouse digestive tract. Toxicon 1992;30:1583–9.
[33] Massot J, Sanchez O, Couchy R, et al. Bacterio–pharmacological activity of *Saccharomyces boulardii* in clindamycin induced colitis in the hamster. Arzneimittelforschung 1984;34:794–7.
[34] Castex F, Corthier G, Joubert S, et al. Prevention of *Clostridium difficile*-induced experimental pseudomembranous colitis by *Saccharomyces boulardii*: a scanning electron microscopic and microbiological study. J Gen Microbiol 1990;136:1085–9.
[35] Czerucka D, Nano JL, Bernasconi P, et al. Réponse aux toxines A et B de *Clostridium difficile* d'une lignée de cellules épithéliales de rat: IRD 98. Effet de *Saccharomyces boulardii*. Gastroenterol Clin Biol 1991;15:22–7.
[36] Elmer GW, Corthier G. Modulation of *Clostridium difficile* induced mortality as a function of the dose and viability of the *Saccharomyces boulardii* used as preventive agent in gnotobiotic mice. Can J Microbiol 1991;37:315–7.
[37] Castagliuolo I, LaMont JT, Nikulasson ST, et al. *Saccharomyces boulardii* protease mediates *Clostridium difficile* toxin A effects in rat ileum. Infect Immun 1996;12:5225–32.
[38] Qamar A, Aboudala S, Warny M. *Saccharomyces boulardii* stimulates intestinal immunoglobulin A immune response to *Clostridium difficile* toxin A in mice. Infect Immun 2001;69:2762–5.
[39] Tasteyre A, Barc MC, Karjalainen T, et al. Inhibition of *in vitro* cell adherence of *Clostridium difficile* by *Saccharomyces boulardii*. Microb Pathog 2002;32:219–25.
[40] Vidon V, Huchet B, Rambau JC. Influence de *Saccharomyces boulardii* sur la sécrétion jéjunale induite chez le rat par la toxine cholérique. Gastroenterol Clin Biol 1986;10:13–6.
[41] Czerucka D, Nano JL, Bernasconi P, et al. Réponse à la toxine cholérique de deux lignées de cellules épithéliales intestinales. Effet de *Saccharomyces boulardii*. Gastroenterol Clin Biol 1989;13:383–7.

[42] Czerucka D, Rampal P. Effect of *Saccharomyces boulardii* on CAMP- and Ca^{2+} dependent Cl^- secretion in T84 cells. Dig Dis Sci 1999;44:2359–68.
[43] Barrett KA. Positive and negative regulation of chloride secretion. Am J Physiol 1993;265:C859–68.
[44] Venglarik CJ, Bridges RJ, Frizel RA. A simple assay for agonist-regulated Cl and K conductances in salt-secreting epithelial cells. Am J Physiol 1990;259:C358–60.
[45] Massot J, Descanclois M, Astoin J. Protection par *Saccharomyces boulardii* de la diarrhée à *Escherichia coli* du souriceau. Ann Pharm Fr 1982;40:445–9.
[46] Philpott PJ, McKay DM, Sherman PM, et al. Infection of T84 cells with enteropathogenic *Escherichia coli* alters barrier and transport functions. Am J Physiol 1998; 33:G634–45.
[47] Savkovic SD, Koutsouris A, Hecht G. Attachment of a non invasive enteric pathogen, enteropathogenic *Escherichia coli*, to cultured human intestinal epithelial monolayers induces transmigration of neutrophils. Infect Immun 1996;64:4480–7.
[48] Yuhan R, Koutsovris A, Savkovic SD, et al. Enteropathogenic *Escherichia coli* induced myosin light chain phosphorylation alters intestinal epithelial permeability. Gastroenterology 1997;113:1873–82.
[49] Dahan S, Dalmasso G, Imbert V, et al. *Saccharomyces boulardii* interferes with enterohemorragic *Escherichia coli* induced signaling pathways in T84 cells. Infect Immun 2003;71:766–73.
[50] Dahan S, Bussatil V, Imbert V, et al. Enterohemorragic *Escherichia coli* infection induces interleukin-8 production via activation of mitogen-activated protein kinases and the transcription factors NF-kappa B and AP-1 in T84 cells. Infect Immun 2002;70:2304–10.
[51] Marandi S, De Keyser N, Stilmant C, et al. Ontogeny of MAP kinases in rat small intestine: premature stimulation by insulin of BBM hydrolases is regulated by ERKs but not by p-38 MAP kinase. Pediatr Res 2002;5(2):180–8.
[52] Marandi S, De Keyser N, Saliez A, et al. Insulin signal transduction in rat small intestine: role of MAP kinases in expression of mucosal hydrolases. Am J Physiol Gastrointest Liver Physiol 2001;280(2):G229–40.
[53] Hobbies S, Chen LM, Davis RJ, et al. Involvement of mitogen-activated protein kinase pathways in the nuclear responses and cytokine production induced by *Salmonella typhimurium* in cultured intestinal epithelial cells. J Immunol 1997;57:1290–8.
[54] Tang P, Sutherland YA, Gold MR, et al. *Listeria* monocytogenes invasion of epithelial cells requires the MEK-1/MEK-2 mitogen-activated protein kinase pathway. Infect Immun 1998;66:1106–12.
[55] Rodrigues ACP, Nardi RM, Bambirra EA, et al. Effect of *Saccharomyces boulardii* against experimental oral infection with *Salmonella typhimurium* and *Shigella flexneri* conventional and gnotobiotic mice. J Appl Bacteriol 1996;91:251–6.
[56] Jahn HU, Ullrich R, Schneider T. Immunology and trophical effects of *Saccharomyces boulardii* on the small intestine in healthy human volunteers. Digestion 1996;57:95–104.
[57] Harms HK, Bertele-Harms RM, Bruer-Kleis D. Enzyme substitution therapy with the yeast *Saccharomyces boulardii* in congenital sucrase-isomaltase deficiency. N Engl J Med 1987;316:1306–9.
[58] Buts JP, De Keyser N, Stilmant C, et al. *Saccharomyces boulardii* enhances N-terminal peptide hydrolysis in suckling rat intestine by endoluminal release of a zinc-binding metalloprotease. Pediatr Res 2002;51:528–34.
[59] Zaouche A, Loukil C, de la Gausie P. Effects of oral *Saccharomyces boulardii* on bacterial overgrowth, translocation and intestinal adaptation after small bowel resection in rats. Scand J Gastroenterol 2000;35:160–5.
[60] Buts JP, De Keyser N, De Raedemaeker L. *Saccharomyces boulardii* enhances rat intestinal enzyme expression by endoluminal release of polyamines. Pediatr Res 1994;36:522–7.
[61] Bowling TF, Raimundo AH, Grimble GK, et al. Reversal by short-chain fatty acids of colonic fluid secretion induced by enteral feeding. Lancet 1993;342:1266–8.

[62] Breves G, Faul K, Schröder B, et al. Application of the colon-stimulation technique for studying the effects of *Saccharomyces boulardii* on basic parameters of porcine cecal microbial metabolism disturbed by clindamycin. Digestion 2000;61:193–200.
[63] Krammer M, Karbach U. Antidiarrheal action of the yeast *Saccharomyces boulardii* in the rat small and large intestine by stimulating chloride absorption. Z Gastrenterol 1993;31:73–7.
[64] Girard P, Pansart Y, Lorette J, et al. Dose–response relationship and mechanism of action of *Saccharomyces boulardii* in castor oil-induced diarrhea in rats. Dig Dis Sci 2003;48: 770–4.
[65] Sougioultzis S, Simeonidis S, Bhaskar KR, et al. *Saccharomyces boulardii* produces soluble anti-inflammatory factor that inhibits NF-KB-mediated IL-8 gene expression. Gastroenterology 2003;125:606.
[66] Kelly CP, Pothoulakis C, LaMont JT. Pseudomembranous colitis. N Engl J Med 1994; 330:257–62.
[67] Benoit R, Dorval D, Loulergue J. Diarrhée postantibiotique: rôle de *Klebsiella oxytoca*. Gastroenterol Clin Biol 1992;16:860–4.
[68] McFarland LV. Risk factors for antibiotic-associated diarrhea. A review of the literature. Ann Med Interne 1998;149:261–6.
[69] Adam J, Barret A, Barret-Ballet C. Essais cliniques contrôlés en double insu de l'Ultra-Levure lyophilisée. Etude multi-centrique par 25 médecins de 388 cas. Med Chir Dig 1976;5: 401–5.
[70] Surawicz CM, McFarland LV, Elmer G, et al. Treatment of recurrent *Clostridium difficile* colitis with vancomycin and *Saccharomyces boulardii*. Am J Gastroenterol 1989; 84:1285–7.
[71] Chapoy P. Traitement des diarrhées aiguës infantiles: essai contrôlé de *Saccharomyces boulardii*. Ann Pediatr 1985;32:561–3.
[72] Tempe JD, Steidel AL, Bléhaut H, et al. Prévention par *Saccharomyces boulardii* des diarrhées de l'alimentation enérale à debit continu. Sem Hop Paris 1983;59:1409–12.
[73] Schlotterer M, Bernasconi P, Lebreton F, et al. Intérêt de *Saccharomyces boulardii* dans la tolérance digestive de la nutrition entérale à débit continu chez le brûlé. Nutrition Clinique Metabolisme 1987;1:31–4.
[74] Saint-Marc T, Bléhaut H, Musial C, et al. Diarrhée en relation avec le SIDA. Essai en double aveugle de *Saccharomyces boulardii*. Sem Hop Paris 1995;71:735–41.
[75] Maupas JL, Champemont P, Delforge M. Traitement des colopathies fonctionnelles. Essai en double aveugle de l'Ultra-Levure (treatment of irritable bowel syndrome with *Saccharomyces boulardii*: a double blind, placebo controlled study). Med Chir Dig 1983;12:77–9.
[76] Hennequin C, Kauffman-Lacroix C, Jobert A. Possible role of catheters in *Saccharomyces boulardii* fungemia. Eur J Clin Microbiol Infect Dis 2000;19:16–20.
[77] Perapoch J, Places AM, Querol A. Fungemia with *Saccharomyces boulardii*, only one of whom had been treated with Ultra-Levure. Eur J Clin Microbiol Infect Dis 2000;19:468–70.

ELSEVIER
SAUNDERS

Gastroenterol Clin N Am 34 (2005) 533–545

GASTROENTEROLOGY CLINICS
OF NORTH AMERICA

The Use of Probiotics in Functional Bowel Disease

Eamonn M.M. Quigley, MD, FRCP, FACP, FACG, FRCPI

Alimentary Pharmabiotic Center, National University of Ireland, Department of Medicine, Cork University Hospital, Clinical Sciences Building, Cork, Ireland

The term functional bowel disease refers to a group of disorders that feature various gastrointestinal (GI) symptoms, such as abdominal pain or discomfort, diarrhea, constipation, bloating, and distension for which there is no obvious organic cause. Foremost among these is irritable bowel syndrome (IBS), a common disorder usually defined by the coexistence of abdominal pain or discomfort and an alteration in bowel habit [1–3]. IBS may lead to impaired social and personal function and can diminish quality of life to a degree usually associated with major organic diseases such as hypertension and diabetes [4,5]. IBS continues to represent a significant therapeutic challenge; currently available therapies provide symptomatic relief, at best, and none have been shown to alter the natural history of the disorder [1,2,6–11].

The precise pathophysiology of IBS remains unknown [12]. For some time, pathophysiological and pharmacological research efforts have focused on two principal targets: dysmotility and altered visceral sensation [13]. There is no doubt that IBS is associated with several disturbances in motor function, not only in the colon, but throughout the GI tract, as reflected by the former use of such terms as spastic colon or spastic colitis to describe this syndrome and by the continued emphasis on the use of antispasmodics in its therapy [11]. Although it remains likely that dysmotility, or spasm, may play a role in the precipitation of symptoms, the specificity of any proposed motor phenomenon for IBS has been questioned, and the primacy of dysmotility in the pathogenesis of IBS is now in doubt [14,15]. Visceral hypersensitivity, usually demonstrated by assessing the response to balloon inflation in the rectum or elsewhere, is a common phenomenon in IBS [16] and in all functional GI disorders, including functional dyspepsia and noncardiac chest pain. Visceral hypersensitivity and related phenomena, such as visceral hyperalgesia and abnormal central perception, are so common in IBS that visceral hypersensitivity, elicited

Supported by grants from Science Foundation Ireland.

E-mail address: e.quigley@ucc.ie

0889-8553/05/$ – see front matter
doi:10.1016/j.gtc.2005.05.008

gastro.theclinics.com

by inflation of a rectal barostat balloon in the rectum, has been proposed as a diagnostic test for IBS [16]. Akin to the situation with dysmotility, the specificity of this phenomenon for IBS has been questioned; some experts in the field contend that visceral hypersensitivity reflects the impact of psychological traits associated with, or consequent upon, IBS and that it is not, therefore, a fundamental pathophysiological mechanism in IBS [17].

More recently, roles for enteric infection and intestinal inflammation have been proposed. Thus, retrospective and prospective studies have documented the new onset of IBS following bacteriologically confirmed bacterial gastroenteritis [18–24], and others have provided evidence of low-grade mucosal inflammation [25–27] and immune activation [28–30] in patients who have IBS. The enteric flora also have been implicated; there has been a suggestion that some patients who have with IBS may harbor bacterial overgrowth and that their symptoms may be ameliorated by its eradication [31–34]. Despite these observations, the ever-increasing understanding of gut flora–mucosa interactions [35] and the existence of a significant body of basic research to support a role for inflammatory and immune processes in contributing to enteric neuromuscular dysfunction [36], the role of lumen–mucosa interactions, in IBS, remains largely unexplored. It is in this context that probiotics have come to be evaluated in the management of IBS.

WHY USE PROBIOTICS IN IRRITABLE BOWEL SYNDROME?

Probiotics, defined as live or attenuated bacteria or bacterial products that confer a significant health benefit to the host [37], have the potential to provide a clinical tool to explore interactions between the resident flora, the intestinal epithelium and the gut-, or mucosa-associated lymphoid tissue (GALT or MALT). There are several reasons why these agents might prove of therapeutic benefit in IBS.

Antibacterial and antiviral effects

Many probiotic organisms exert antibacterial and antiviral effects and could, thereby, prevent or modify the course of postinfective IBS [38–44]. Probiotics have been shown to be beneficial for preventing such human diarrheal conditions as toddlers' diarrhea and *Clostridium difficile*-related, antibiotic-associated diarrhea [38,39,45,46]. These effects could be especially relevant to postinfectious IBS. Real data are beginning to emerge that directly support the concept of postinfective or postdysenteric IBS. Just a few years ago, in a retrospective analysis, McKendrick and Read reported on 12 patients (out of 38) who developed chronic bowel dysfunction within 12 months of documented *Salmonella* gastroenteritis [18]. Taking advantage of the presence of centralized public health microbiology laboratories in the United Kingdom, Neal and colleagues performed a retrospective analysis on a large group of patients who had documented bacterial gastroenteritis [19]. When followed 6 months later, 7% had developed symptoms consistent with IBS. They went on to identify the following risk factors for postinfectious IBS: being female and having a prolonged initial illness. Another study, also from the United Kingdom, included 318 patients with gastroenteritis followed for 1 year [20]. This study included a control population drawn from a national database. The rate of new diagnoses of IBS during the 12-month period of study was 4.4% for those with prior exposure to gastroenteritis and 0.3% for the control group, with a relative risk for developing IBS after gastroenteritis in excess of 10. Recent prospective studies have clarified the concept of post-infectious IBS further. The first included 75 patients with acute gastroenteritis who were followed for at least 6 months [21]. At 3 months, 22 had developed IBS. Similar risk factors were identified: female gender and a prolonged episode of gastroenteritis. In addition, they noted that the patients who developed IBS had higher scores for anxiety, depression, somatization, and neurosis. They also provided some insights into the natural history of this disorder. At 6 months, 91% had persistent symptoms; at 9 months 79% had persistent symptoms, and at 12 months 75% had persistent symptoms, suggesting that many of these individuals truly had developed chronic IBS. A second study from the same group evaluated 94 patients with acute gastroenteritis, and a control group, at the onset of the symptoms of gastroenteritis and 3 months later with a questionnaire, psychometric testing, flexible sigmoidoscopy (with biopsy), whole gut transit study, and rectal distention [47]. They carefully excluded all patients with prior GI dysfunction, something that had not always been done in other studies. At 3 months, 23% of the gastroenteritis group had developed IBS. Again, 64% of those who developed postinfectious IBS were female. Psycho–social factors again proved predictive of IBS risks. Recent experience of a major life event and a high hypochondriasis score were especially predictive for developing postinfectious IBS. For the first time, this study explored the possible role of inflammation. At the onset of gastroenteritis, 37% had evidence of microscopic colitis, regardless of whether they subsequently went on to develop IBS. At 3 months, in contrast, an increase in the number of chronic inflammatory cells was seen only among the patients who had developed IBS. In a very recent study, another group demonstrated a persisting increase in rectal mucosal enteroendocrine cells, T lymphocytes and in gut permeability in patients who had postdysenteric IBS [22]. Finally, an innovative, intriguing study surveyed individuals about to travel abroad to high-risk areas for enteric infections before travel, immediately on return, and at least 12 weeks later [48]. Forty-six percent developed traveler's diarrhea. At 12 weeks, 9.5% had new-onset IBS by Rome criteria, a relative risk of about 7, very consistent with other studies. Although probiotics have not been evaluated in the context of postinfectious IBS, the ability of probiotic preparations to influence the outcome of bacterial infections, such as *C difficile* colitis, and viral infections, such as rotavirus diarrhea, and the experimental demonstration of bacteriocidal, toxin-neutralizing and antiviral effects for specific probiotic strains, suggest that probiotics may have a role for preventing or treating postinfectious IBS.

Anti-inflammatory effects

IBS also may be associated with inflammation or immune activation in the absence of an infectious trigger. That inflammation could lead to altered enteric nerve or muscle function had been demonstrated in the past in such disorders as Chagas' disease and postviral gastroparesis. IBS-type symptoms also have been associated with inflammatory bowel disease (IBD) [49] and celiac disease [50], even when in apparent remission. More direct and compelling evidence recently was provided by Chadwick and colleagues, who evaluated 77 patients who had IBS, of whom 55% would be considered as diarrhea-predominant; none had a confirmed infectious origin for their IBS [26]. All had colonic biopsies taken for conventional histology and immunohistology. Thirty-eight had normal histology; 31 demonstrated microscopic inflammation, and eight fulfilled criteria for lymphocytic colitis. Among the group with normal histology, however, immunohistology revealed increased intraepithelial lymphocytes, and an increase in CD3+ and CD25+ cells in the lamina propria. All, therefore, showed evidence of immune activation. These features were even more evident in the microscopic inflammation group. These patients additionally revealed increased neutrophils, mast cells, and natural killer cells. All of these immunopathological abnormalities were most evident in the lymphocytic colitis group. These patients also demonstrated HLA-DR staining in crypts and increased CD8+ cells in the lamina propria. Taking the group of IBS patients as a whole, CD3+ cell numbers were higher among those with diarrhea than among alternators or those with predominant constipation. In contrast, among the noninflamed IBS group, the presence of mast cells was a predictor of constipation. Surprisingly, given the described direct relationship between symptoms and chronic inflammation among patients who have postinfectious IBS, these authors did not find an association between the nature of disease onset or disease duration and immunological findings. That patients who have IBS may be predisposed to an inflammatory response to luminal triggers is also supported by the finding of a reduced frequency of the high-producer phenotype for the anti-inflammatory cytokine interleukin (IL)-10 among patients who have IBS [29]. Most recently, among a group of 78 unselected IBS patients, the authors demonstrated in peripheral blood mononuclear cells an alteration in the ratio between the cytokines IL-10 and IL-12. This became skewed toward a Th1, proinflammatory profile [28].

Although the inflammatory hypothesis in IBS is in its infancy, there is already some evidence for the extension of the inflammatory process beyond the confines of the mucosal compartment. Tornblom and colleagues addressed this issue in 10 patients who had severe IBS by examining full-thickness jejunal biopsies obtained at laparoscopy [27]. In nine patients, they found low-grade infiltration of lymphocytes in the myenteric plexus; four of these had an associated increase in intraepithelial lymphocytes, and six patients demonstrated evidence of neuronal degeneration. Nine patients had longitudinal muscle hypertrophy, and seven patients had abnormalities in the number and size of interstitial cells of Cajal. Three of their patients reported an acute onset of their IBS; in two patients, this possibly was precipitated by gastroenteritis. The finding of intraepithelial lymphocytosis is consistent with the reports of Chadwick and colleagues [26] in the colon and of Wahnschaffe and colleagues in the duodenum [51].

A direct linkage between immune activation and symptoms has been provided by the work of Barbara and colleagues, who demonstrated, not only an increased prevalence of mast cell degranulation in the colon in IBS, but also a direct correlation between the proximity of mast cells to neuronal elements and pain severity [52].

What is the pathogenesis of these inflammatory changes in IBS? With regard to postinfectious IBS, Spiller proposed that the development of chronic inflammation could represent a response to an initial bacterial infection among individuals rendered susceptible by a relative deficiency of anti-inflammatory cytokines [53]. A similar hypothesis has been advanced to explain inflammation and immune activation in IBS, in general, where a failure to adequately down-regulate a proinflammatory response following a precipitating event (eg, GI infection) may sustain the IBS state [28,29]. Why than does this inflammatory response not progress to the state of florid inflammation so characteristic of IBD? In this regard, Collins suggests that the prominence of CD25+ cells in the inflammatory infiltrate, as reported by Chadwick and colleagues, may act to prevent progression to a more florid inflammatory response [25].

Laboratory experiments repeatedly have demonstrated the anti-inflammatory effects of certain probiotics. For example, oral administration of a *Bifidobacterium* exerted a profound anti-inflammatory effect in the IL-10 knock-out mouse, a potent model of IBD that was associated with suppression of the proinflammatory cytokines IFNγ, tumor necrosis factor (TNF)-α, and IL-12, while preserving activity of the anti-inflammatory cytokine TGFβ [54,55]. Others have demonstrated similar effects for the probiotic cocktail VSL#3 in another animal model of colitis; in this instance, the anti-inflammatory effect was evident with bacterial DNA alone [56]. In clinical practice, probiotics have been demonstrated to prevent development of pouchitis [57] and reduce the relapse rate of this condition following successful antibiotic treatment [58,59]. By reducing mucosal inflammation [54–60], probiotics could decrease immune-mediated activation of enteric motor and sensory neurons and modify neural traffic between the gut and the central nervous system. There is some preliminary evidence that probiotic administration may diminish visceral hypersensitivity in animal models [61,62]. Furthermore, effects on motility and perception could go some way toward explaining the beneficial effects of probiotics on bloating, given current concepts on the roles of altered gas transit and visceral hypersensitivity in the pathogenesis of this symptom [63–65].

Most recently, a probiotic bacterium, *Lactobacillus paracasei,* has been shown to reverse changes in intestinal muscle function induced by inflammation consequent upon infection of an animal model with *Trichinella spiralis* [66]. This latter finding indicates that a luminally administered probiotic can influence inflammatory processes beyond the mucosal surface.

In the author's recent study, which compared the effects of two probiotics, a *Lactobacillus* and a *Bifidobacterium,* in IBS, efficacy was observed with the *Bifidobacterium* alone. Bifidobacterium, but not the *Lactobacillus* or placebo, normalized the disturbed cytokine ratio, providing, for the first time in people, evidence for an association between an effect on an inflammatory process and amelioration of symptoms [28].

Altering the composition of the gut flora

Although the status of the gut flora in IBS remains a source of some controversy [12,67,68], probiotic-related changes in the enteric flora could influence gut function either directly, through the augmentation of commensal lactobacilli or bifidobacteria or the elimination of pathogens, or indirectly, through a reduction in pathogen-related inflammation [60] or bacterial fermentation [69]. Whether IBS is accompanied by quantitative or qualitative changes in the bacterial flora of the small or large intestine remains a contentious issue. Although some have described bacterial overgrowth in the small intestine [32–33] and qualitative alterations in the fecal flora [67,68,70] and increased bacterial fermentation [69], others have failed to replicate these findings [12]. A reduction in bacterial fermentation by a modulation of the composition of the flora could contribute to the alleviation of the gas-related symptoms common in IBS [63,64] and that appear to reflect a selective defect in intestinal gas transport [65]. An effect on bloating may be the most consistent effect of probiotics across all studies.

Effects on luminal contents

Probiotics could alter the volume or composition of stool and gas [71] or increase intestinal mucus secretion [72]. These effects could influence intestinal handling and thus modulate such symptoms as constipation and diarrhea. Of these putative effects of probiotics, an effect on stool bulking would appear unlikely, given the findings of the author's study of a *Bifidobacterium* in IBS [28]. In contrast to the effects observed with two newly approved therapies for IBS, namely alosetron [73] and tegaserod [74,75], the author failed to observe any change in stool consistency or frequency. This apparent independence of the effects of *Bifidobacterium* from any change in stool frequency or form implies that this therapeutic approach may be applicable to all patients who have IBS, irrespective of stool pattern.

Effects on gut flora and luminal contents are not mutually exclusive and could interact with other factors known to influence symptom onset in IBS. As illustrated in Fig. 1, bacterial overgrowth or qualitative changes in the gut flora with a shift toward more gas-producing organisms could interact with unabsorbed carbohydrates (such as in the patient with lactase deficiency or fructose intolerance) to increase colonic fermentation, which could, not only increase intestinal gas-related symptoms, but also affect function in the proximal gut by promoting gastroesophageal reflux [76] and modifying proximal gastric relaxation [77]. These latter effects could contribute to the

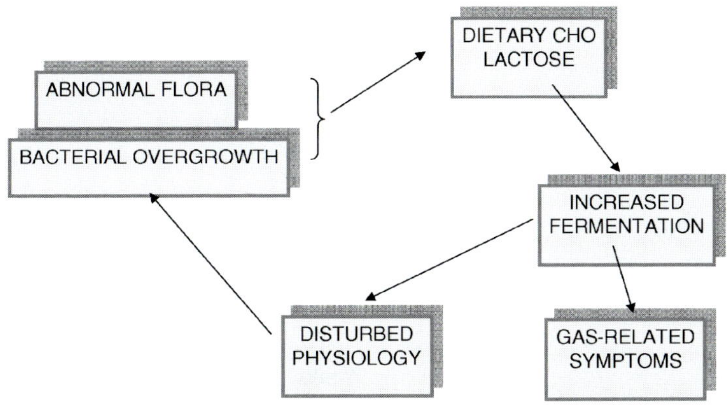

Fig. 1. Potential inter-relationships between the bacterial flora, dietary constituents, gut function, and symptoms in functional disorders such as IBS.

overlap between IBS and other functional GI disorders such as nonerosive gastroesophageal reflux disease and functional dyspepsia.

Studies of probiotics in irritable bowel syndrome

Numerous studies have evaluated the response of IBS to probiotic preparations and, while results between studies are difficult to compare because of differences in study design, probiotic dose, and strain, there has been some evidence of symptom improvement [78–87]. The overall impact of probiotics, in IBS, remains unclear [88–92]. In a recent review, Hamilton-Miller, while drawing attention to the shortcomings of prior trials in terms of study design, concluded that there was overall sufficient evidence of efficacy to warrant further evaluation [89]. Most studies reviewed were small and almost certainly underpowered to demonstrate anything other than a very striking benefit. Several did not verify bacterial transit and survival by confirmatory stool studies. Many different organisms and strains were employed, and dosage varied from as little as 10^5 to 10^{13}. Furthermore, some, including a very recent study, employed probiotic cocktails rather than single isolates, rendering it impossible to induce what were the active moieties [87]. Nevertheless, some positive results were noted. Niedzielin reported resolution of abdominal pain in all 20 patients treated for 4 weeks with *L plantarum* 299V, in contrast to only 11 of 20 patients who received a placebo [82]. Halpern and colleagues noted a significant reduction in an IBS symptom index with a capsule containing 5×10^9 heat-killed *L acidophilus* [78]. O'Sullivan and colleagues, while failing to detect an effect of *L casei* GG on overall symptomatology, did note a trend toward reduction in bloating [80]. Nobaek and colleagues, employing *L plantarum* [79], described a similar benefit in terms of relief of bloating, as did Kim and colleagues in their evaluation of VSL#3 [87].

The author's study compared, for the first time, the effects of two probiotic strains on symptoms in patients who had IBS [28]. The author's group

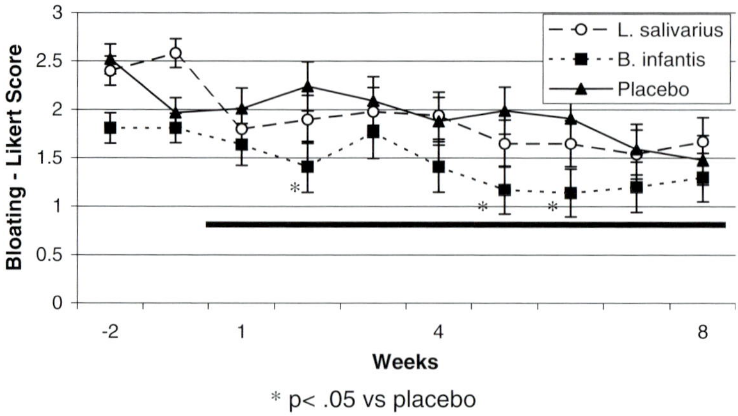

Fig. 2. Comparative effects of a *Bifidobacterium*, a *Lactobacillus*, and placebo and bloating in IBS over an 8-week treatment period (indicated by the solid line).

demonstrated superiority for a *B longum infantis* over *Lactobacillus* or placebo for each of the cardinal symptoms of the IBS (abdominal pain/discomfort, distension/bloating, and difficult defecation), and for a composite score. For each individual symptom, with the notable exceptions of bowel movement frequency and consistency, the group randomized to *B infantis* experienced a greater reduction in symptom scores during the treatment period. Fig. 2, for instance, compares bloating scores for the three treatment groups. These symptomatic benefits were associated with parallel trends in quality of life. Furthermore, this therapy was tolerated well and free of significant adverse events. As these benefits were observed independent of any change in stool frequency or form, they cannot be attributed to either a laxative or an antidiarrheal effect. Although this study did not involve a comparison with any other therapeutic modality, and the study design differed, in some aspects, from recent large trials of serotonergic agonists and antagonists, the therapeutic gain observed for *Bifidobacterium* over placebo (20% to 25%) was certainly no less than that reported for tegaserod and alosetron (10% to 20%) [73–75]. This study with *B longum* therefore provides clear evidence for a benefit, in IBS, for a clearly defined single-organism probiotic preparation and suggests that some strains may be more effective that others for this indication. This was a relatively small pilot study, and its findings must be interpreted with caution. Large, randomized controlled trials of this *Bifidobacterium* strain, however, are warranted in IBS, and detailed explorations of its mechanism(s) of action are indicated.

SUMMARY

The clear delineation of a postinfective variety of IBS, and the description in numerous studies of evidence of low-grade inflammation and immune activation in IBS, suggest a role for a dysfunctional relationship between the indigenous flora and the host in IBS. Accordingly, this provides a clear rationale for the use of probiotics in this disorder. Other modes of action, including bacterial displacement and alterations in luminal contents, are also plausible. Clinical evidence of efficacy remains scanty, and a review of available trials, while providing some hints of efficacy and therapeutic promise, emphasizes the importance of clear definition of strain selection, dose, and viability. This is clearly an area of great potential in IBS and deserves further study at all levels.

References

[1] Brandt LJ, Locke GR, Olden K, et al. An evidence-based approach to the management of irritable bowel syndrome in North America. AJG 2002;97:S1–26.
[2] Drossman DA, Camilleri M, Mayer EA, et al. AGA technical review on irritable bowel syndrome. Gastroenterology 2002;123:2108–31.
[3] Thompson WG, Longstreth GF, Drossman DA, et al. Functional bowel disorders and functional abdominal pain. Gut 1999;45(Suppl 2):S43–7.
[4] Thompson WG, Heaton KW, Smyth GT, et al. Irritable bowel syndrome in general practice: prevalence, characteristics, and referral. Gut 2000;46:78–82.
[5] Hungin APS, Whorwell PJ, Tack J, et al. The prevalence, patterns and impact of irritable bowel syndrome: an international survey of 40,000 subjects. Aliment Pharmacol Ther 2003;17:643–50.
[6] Jailwala J, Imperiale TF, Kroenke K. Pharmacologic treatment of the irritable bowel syndrome: a systematic review of randomized, controlled trials. Ann Intern Med 2000;133:136–47.
[7] Jones J, Boorman J, Cann P, et al. British Society of Gastroenterology guidelines for the management of the irritable bowel syndrome. Gut 2000;47(Suppl 2):ii1–19.
[8] Akehurst R, Kaltenthaler E. Treatment of irritable bowel syndrome: a review of randomised controlled trials. Gut 2001;48:272–82.
[9] Camilleri M. Management of the irritable bowel syndrome. Gastroenterology 2001;120:652–68.
[10] Talley NJ. Pharmacologic therapy for the irritable bowel syndrome. Am J Gastroenterol 2003;98:750–8.
[11] Poynard T, Naveau S, Mory B, et al. Meta-analysis of smooth muscle relaxants in the treatment of irritable bowel syndrome. Aliment Pharmacol Ther 1994;8:499–510.
[12] Quigley EMM. Current concepts of the irritable bowel syndrome. Scand J Gastroenterol 2003;38(suppl 237):1–8.
[13] Mayer EA, Collins SM. Evolving pathophysiological models of functional gastrointestinal disorders. Gastroenterology 2002;122:2032–48.
[14] McKee DP, Quigley EMM. Intestinal motility and the irritable bowel syndrome—is IBS a motility disorder? Part 1. Definition of IBS and colonic motility. Dig Dis Sci 1993;38:1761–72.
[15] McKee DP, Quigley EMM. Intestinal motility and the irritable bowel syndrome—is IBS a motility disorder? Part 2. Motility of the small bowel, esophagus, stomach and gall bladder. Dig Dis Sci 1993;38:1773–82.
[16] Bouin M, Plourde V, Boivin M, et al. Rectal distension testing in patients with irritable bowel syndrome: sensitivity, specificity and positive predictive values of pain sensory thresholds. Gastroenterology 2002;122:1771–7.
[17] Whitehead WE, Palsson OS. Is rectal pain sensitivity a biological marker for irritable bowel syndrome: psychological influences on pain perception. Gastroenterology 1998;115:1263–71.

[18] McKendrick MW, Read MW. Irritable bowel syndrome—post-*Salmonella* infection. J Infect 1994;29:1–3.
[19] Neal KR, Hebdon J, Spiller R. Prevalence of gastrointestinal symptoms six months after bacterial gastroenteritis and risk factors for development of the irritable bowel syndrome. BMJ 1997;314:779–82.
[20] Garcia Rodriguez LA, Ruigomez A. Increased risk of irritable bowel syndrome after bacterial gastroenteritis: cohort study. BMJ 1999;318:565–6.
[21] Gwee KA, Graham JC, McKendrick MW, et al. Psychometric scores and persistence of irritable bowel after infectious diarrhoea. Lancet 1996;347:150–3.
[22] Spiller RC, Jenkins D, Thornley JP, et al. Increased rectal mucosal enteroendocrine cells, T lymphocytes, and increased gut permeability following acute *Campylobacter enteritis* and in post-dysenteric irritable bowel syndrome. Gut 2000;47:804–11.
[23] Dunlop SP, Jenkins D, Neal KR, et al. Relative importance of enterochromaffin cell hyperplasia, anxiety, and depression in postinfectious IBS. Gastroenterology 2003; 125:1651–9.
[24] Cumberland P, Sethi D, Roderick PJ, et al. IID Study Executive. The infectious intestinal disease study of England: a prospective evaluation of symptoms and health care use after an acute episode. Epidemiol Infect 2003;130:453–60.
[25] Collins SM. A case for an immunological basis for irritable bowel syndrome. Gastroenterology 2002;122:2078–80.
[26] Chadwick V, Chen W, Shu D, et al. Activation of the mucosal immune system in irritable bowel syndrome. Gastroenterology 2002;122:1778–83.
[27] Tornblom H, Lindberg G, Nyberg B, et al. Full-thickness biopsy of the jejunum reveals inflammation and enteric neuropathy in irritable bowel syndrome. Gastroenterology 2002;123:1972–9.
[28] O'Mahony L, McCarthy J, Kelly P, et al. *Lactobacillus* and *Bifidobacterium* in irritable bowel syndrome: symptom responses and relationship to cytokine profiles. Gastroenterology 2005;128:541–51.
[29] Gonsalkorale WM, Perrey C, Pravica V, et al. Interleukin 10 genotypes in irritable bowel syndrome: evidence for an inflammatory component? Gut 2003;52:91–3.
[30] Gwee K-A, Collins SM, Read NW, et al. Increased rectal mucosal expression of interleukin 1beta in recently acquired postinfectious irritable bowel syndrome. Gut 2003;52:523–6.
[31] Pimentel M, Chow EJ, Lin HC. Eradication of small bowel bacterial overgrowth reduces symptoms of irritable bowel syndrome. Am J Gastroenterol 2000;95:3503–6.
[32] Pimentel M, Chow E, Lin H. Normalization of lactulose breath testing correlates with symptom improvement in irritable bowel syndrome: a double-blind, randomized, placebo-controlled study. Am J Gastroenterol 2003;98:412–9.
[33] Tursi A, Brandimarte G, Giorgetti GM. High prevalence of small intestinal bacterial overgrowth in celiac patients with persistence of gastrointestinal symptoms after gluten withdrawal. Am J Gastroenterol 2003;98:839–43.
[34] O'Leary C, Quigley EMM. Small bowel bacterial overgrowth, celiac disease and IBS: what are the associations? Am J Gastroenterol 2003;98:720–2.
[35] Shanahan F. Therapeutic manipulation in gut flora. Science 2000;289:1311–2.
[36] Collins SM. The immunomodulation of enteric neuromuscular function: implications for motility and inflammatory disorders. Gastroenterology 1996;111:1683–99.
[37] Gorbach SL. Probiotics in the third millennium. Dig Liver Dis 2002;34(Suppl 2):S2–7.
[38] Isolauri E, Kirjavainen PV, Salminen S. Probiotics: a role in the treatment of intestinal infection and inflammation? Gut 2002;50:iii54–9.
[39] Perdigon G, Alvarez S, Nader DE, et al. The oral administration of lactic acid bacteria increases the mucosal immunity in response to enteropathogens. J Food Prot 1990;53: 404–40.
[40] Rolfe RD. The role of probiotic cultures in the control of gastrointestinal health. J Nutr 2000;130:396S–402S.

[41] Duffy LC, Leavens A, Griffiths E, et al. Perspectives on bifidobacteria as biotherapeutic agents in gastrointestinal health. Dig Dis Sci 1999;44:1499–505.
[42] Marteau P, Seksisk P, Jian R. Probiotics and intestinal health effects: a clinical perspective. Br J Nutr 2002;88(Suppl 1):S51–7.
[43] Fooks LJ, Gibson GR. Probiotics as modulators of the gut flora. Br J Nutr 2002; 88(Suppl 1):S39–49.
[44] Von Wright A, Salminen S. Probiotics: established effects and open questions. Eur J Gastroenterol Hepatol 1999;11:1195–8.
[45] Allen SJ, Okoko B, Martinez E, et al. Probiotics for treating infectious diarrhoea. Cochrane Database Syst Rev 2003; DOI:10.1002/14651858.CD003048.pub2.
[46] Cremonini F, Di Caro S, Santarelli L, et al. Probiotics in antibiotic-associated diarrhoea. Dig Liver Dis 2002;34:S78–80.
[47] Gwee KA, Leong YL, Graham C, et al. The role of psychological and biological factors in postinfective gut dysfunction. Gut 1999;44:400–6.
[48] Ilnyckyj A, Balachandra B, Elliott L, et al. Post-traveler's diarrhea irritable bowel syndrome: a prospective study. Am J Gastroenterol 2003;98:596–9.
[49] Simren M, Axelsson J, Gillberg R, et al. Quality of life in inflammatory bowel disease in remission: the impact of IBS-like symptoms and associated psychological factors. Am J Gastroenterol 2002;97:389–96.
[50] O'Leary C, Wieneke P, Buckley S, et al. Celiac disease and irritable bowel-type symptoms. Am J Gastroenterol 2002;97:1463–7.
[51] Wahnschaffe U, Ullrich R, Riecken EO, et al. Celiac-like abnormalities in a subgroup of patients with irritable bowel syndrome. Gastroenterology 2001;121:1329–38.
[52] Barbara G, Stanghellini V, De Giorgio R, et al. Activated mast cells in proximity to colonic nerves correlate with abdominal pain in irritable bowel syndrome. Gastroenterology 2004;126:693–702.
[53] Spiller RC. Role of nerves in enteric infection. Gut 2002;51:759–62.
[54] O'Mahony L, Feeney M, O'Halloran S, et al. Probiotic impact on microbial flora, inflammation, and tumour development in IL-10 knockout mice. Aliment Pharmacol Ther 2001;15:1219–25.
[55] McCarthy J, O'Mahony L, O'Callaghan L, et al. Double-blind, placebo-controlled trial of two probiotic strains in interleukin 10 knockout mice and mechanistic link with cytokine balance. Gut 2003;52:975–80.
[56] Rachmilewitz D, Katakura K, Karmeli F, et al. Toll-like receptor 9 signaling mediates the anti-inflammatory effects of probiotics in murine experimental colitis. Gastroenterology 2004;126:520–8.
[57] Gionchetti P, Rizzello F, Helwig U, et al. Prophylaxis of pouchitis onset with probiotic therapy: a double-blind, placebo-controlled trial. Gastroenterology 2003;124: 1202–09.
[58] Gionchetti P, Rizzello F, Venturi A, et al. Oral bacteriotherapy as maintenance treatment in patients with chronic pouchitis: a double-blind, placebo-controlled trial. Gastroenterology 2000;119:305–9.
[59] Mimura T, Rizzello F, Helwig U, et al. Once-daily high-dose probiotic therapy (VSL#3) for maintaining remission in recurrent or refractory pouchitis. Gut 2004;53:108–14.
[60] Jijon H, Backer J, Diaz H, et al. DNA from probiotic bacteria modulates murine and human epithelial and immune function. Gastroenterology 2004;126:1358–73.
[61] Greenwood-van Meerveld B, Johnson AC, et al. Probiotic bacteria normalize post-inflammatory visceral hyperalgesia in rats. Gastroenterology 2002;122:A476.
[62] Lamine F, Cauquil E, Eutamene H, et al. *Lactobacillus farciminis* treatment reduces sensitivity to rectal distension in rats: involvement of nitric oxide. Gastroenterology 2003;122:A476.
[63] Quigley EMM. The role of gas in IBS. In: Camilleri M, Spiller RC, editors. Irritable Bowel Syndrome. Diagnosis and treatment. Philadelphia: Philadelphia: WB Saunders Company; 1992. p. 77–84.

[64] Quigley EMM. From comic relief to real understanding; how intestinal gas causes symptoms. Gut 2003;52:1659–61.

[65] Serra J, Azpiroz F, Malagelada JR. Impaired transit and tolerance of gas in the irritable bowel syndrome. Gut 2001;48:14–9.

[66] Verdu EF, Bercik P, Bergonzelli GE, et al. *Lactobacillus paracasei* normalizes muscle hypercontractility in a murine model of postinfective gut dysfunction. Gastroenterology 2004;127:826–37.

[67] Bradley HK, Wyatt GM, Bayliss CE, et al. Instability in the faecal flora of a patient suffering from food-related irritable bowel syndrome. J Med Microbiol 1987;23:29–32.

[68] Balsari A, Ceccarelli A, Dubini F, et al. The fecal microbial population in the irritable bowel syndrome. Microbiologica 1982;5:185–94.

[69] King TS, Elia M, Hunter JO. Abnormal colonic fermentation in irritable bowel syndrome. Lancet 1998;352:1187–9.

[70] Si JM, Yu YC, Fan YJ, et al. Intestinal microecology and quality of life in irritable bowel syndrome patients. World J Gastroenterol 2004;10:1802–5.

[71] Jiang T, Savaiano DA. Modification of colonic fermentation by bifidobacteria and pH in vitro: impact on lactose metabolism, short-chain fatty acid, and lactate production. Dig Dis Sci 1997;42:2370–7.

[72] Ouwehand AC, Lagstrom H, Suomalainen T, et al. Effect of probiotics on constipation, fecal azoreductase activity and fecal mucin content in the elderly. Ann Nutr Metab 2002;46:159–62.

[73] Cremonini F, Delgado-Aros S, Camilleri M. Efficacy of alosetron in irritable bowel syndrome: a meta-analysis of randomized controlled trials. Neurogastroenterol Motil 2003;15:79–86.

[74] Muller-Lissner SA, Fumagalli I, Bardhan KD, et al. Tegaserod, a 5-HT receptor partial agonist, relieves symptoms in irritable bowel syndrome patients with abdominal pain, bloating and constipation. Aliment Pharmacol Ther 2001;15:1655–66.

[75] Kellow J, Lee OY, Chang FY, et al. An Asia-Pacific, double-blind, placebo-controlled, randomised study to evaluate the efficacy, safety, and tolerability of tegaserod in patients with irritable bowel syndrome. Gut 2003;52:671–6.

[76] Piche T, des Varannes SB, Sacher-Huvelin S, et al. Colonic fermentation influences lower esophageal sphincter function in gastroesophageal reflux disease. Gastroenterology 2003;124:894–902.

[77] Ropert A, Cherbut C, Roze C, et al. Colonic fermentation and proximal gastric tone in humans. Gastroenterology 1996;111:289–96.

[78] Halpern GM, Prindiville T, Blankenburg M, et al. Treatment of irritable bowel syndrome with lacteol fort: a randomized, double-blind, cross-over trial. Am J Gastroenterol 1996; 91:1579–85.

[79] Nobaek S, Johansson ML, Molin G, et al. Alteration of intestinal microflora is associated with reduction in abdominal bloating and pain in patients with irritable bowel syndrome. Am J Gastroenterol 2000;95:1231–8.

[80] O'Sullivan MA, O'Morain CA. Bacterial supplementation in the irritable bowel syndrome. A randomised double-blind placebo-controlled crossover study. Dig Liver Dis 2000; 32:302–4.

[81] Brigidi P, Vitali B, Swennen E, et al. Effects of probiotic administration upon the composition and enzymatic activity of human fecal microbiota in patients with irritable bowel syndrome or functional diarrhea. Res Microbiol 2001;152:735–41.

[82] Niedzielin K, Kordecki H, Birkenfeld B. A controlled, double-blind, randomized study on the efficacy of *Lactobacillus plantarum* 299V in patients with irritable bowel syndrome. Eur J Gastroenterol Hepatol 2001;13:1143–7.

[83] Sen S, Mullan MM, Parker TJ, et al. Effect of *Lactobacillus plantarum* 299v on colonic fermentation and symptoms of irritable bowel syndrome. Dig Dis Sci 2002;47:2615–20.

[84] Bazzocchi G, Gionchetti P, Almerigi PF, et al. Intestinal microflora and oral bacteriotherapy in irritable bowel syndrome. Dig Liver Dis 2002;34(Suppl 2):S48–53.
[85] Adler SN, Jacob H, Eliakim R. The probiotic agent *E coli* strain ATCC20226 has a healing effect on proximal inflammation of the small bowel. Gastroenterology 2002;122:A527.
[86] Parker P, McNaught CE, Anderson ADG, et al. Synbiotic in irritable bowel syndrome: a double-blind prospective randomised controlled trial. Gut 2003;52:A11.
[87] Kim HJ, Camilleri M, McKinzie S, et al. A randomized controlled trial of a probiotic, VSL#3, on gut transit and symptoms in diarrhoea-predominant irritable bowel syndrome. Aliment Pharmacol Ther 2003;17:895–904.
[88] Barbara G, Corinaldesi R. Probiotics: could they turn out to be ineffective in irritable bowel syndrome? Dig Liver Dis 2000;32:294–301.
[89] Hamilton-Miller JMT. Probiotics in the management of irritable bowel syndrome: a review of clinical trials. Microb Ecol Health Dis 2001;13:212–6.
[90] Thompson WG. Probiotics for irritable bowel syndrome: a light in the darkness? Eur J Gastroenterol Hepatol 2001;13:1135–6.
[91] Madden JA, Hunter JO. A review of the role of the gut microflora in irritable bowel syndrome and the effects of probiotics. Br J Nutr 2002;88(Suppl 1):S67–72.
[92] Floch MH. Probiotics, irritable bowel syndrome, and inflammatory bowel disease. Curr Treat Options Gastroenterol 2003;6:283–8.

ELSEVIER
SAUNDERS

Gastroenterol Clin N Am 34 (2005) 547–570

GASTROENTEROLOGY CLINICS
OF NORTH AMERICA

Use of Probiotics in Humans: An Analysis of the Literature

Martin H. Floch, MS, MD[a,b,*], David C. Montrose, BA[c]

[a]Section of Diagnostic Diseases, Yale University School of Medicine, 333 Cedar Street, New Haven, CT 05620, USA
[b]Norwalk Hospital, 30 Stevens Street, Suite E, Norwalk, CT 06850, USA
[c]Center for Molecular Medicine, University of Connecticut Health Center, Farmington, CT 06030, USA

Probiotics are live microbial organisms that are administered in supplements to benefit the host [1,2]. They are taken by patients, administered by health care professionals, and prescribed by physicians [3,4]. This widespread use has not been analyzed clearly, however. The authors reviewed the literature between 1980 and 2004 and found 185 references in credible journals [5]. In analyzing that data, they found 16 organisms were used either in a single probiotic supplement or in combination [5]. In that analysis of the literature, the authors also found that there were 20 types of studies performed (Box 1). Of those, only 16 were open trials, or 8.6%, whereas 164, or 91%, were some form of controlled study [5].

When reviewing the 185 manuscripts, the authors also found that there were 68 clinical conditions evaluated in the 185 studies. This article analyzes the most common clinical studies evaluated and also attempts to add to those studies any meta analysis or significant analysis done by other authors.

PROBIOTICS USED TO TREAT INFECTIONS

Diarrhea

The most common clinical entity studied and reported in the probiotic literature from 1980 to 2004 was diarrhea. Forty one, or 22%, of the 185 studies focused on this clinical problem. Twenty of the 41 studies were performed on children. Thirteen were associated with antibiotic-induced diarrhea, whereas three were studies on patients with radiation diarrhea.

Use in adults with diarrhea

The seven studies reporting on the use of probiotics in adults with diarrhea are listed in Table 1. *Lactobacillus reuteri* [6], *Saccharomyces boulardii* [7–8], *LGG* [9], *Lactobacillus casei* [10], and a combination of *L casei, Streptococcus thermophilus*, and

*Corresponding author. *E-mail address*: martinfloch@snet.net (M.H. Floch).

0889-8553/05/$ – see front matter
doi:10.1016/j.gtc.2005.05.004

gastro.theclinics.com

Box 1: Types of studies (order of prevalence)*

Randomized double-blind placebo-controlled: 79
Randomized placebo-controlled: 28
Randomized controlled: 16
Open trial: 16
Randomized double-blind controlled: 10
Placebo-controlled: 6
Randomized double-blind placebo controlled-crossover: 5
Controlled: 5
Double-blind placebo-controlled: 3
Randomized-blind controlled: 3
Randomized: 2
Randomized triple-blind placebo controlled: 1
Randomized blinded placebo-controlled: 1
Randomized double-blind double dummy-controlled: 1
Randomized double dummy controlled: 1
Double-blind placebo-controlled-crossover: 1
Randomized placebo controlled-crossover: 1
Triple-blind placebo-controlled: 1
Randomized blinded controlled crossover: 1
Randomized double-blind controlled crossover: 1

*Studies used in 185 manuscripts published on probiotic use in people from 1980 to 2004 [5].

Bifidobacterium bulgaricus [11] were used in the six studies. One study using *L acidophilus* had variable results [12]. There was range of doses, and those are included in the table. The authors believe the number of patients is too small, and there are too many dose and probiotic organism variables to perform a meta-analysis, but six of the authors in six studies felt it improved the diarrhea by either shortening the course or decreasing the severity.

Use in children with diarrhea

Most studies are in children and recorded in Table 2. Once again, there is wide range of probiotics, doses, and protocol administration used. *L acidophilus, L reuteri, L rhamnosus, B bifidum*, or a combination of *Bifidobacterium* plus *L acidophilus, L acidophilus* plus *L bulgaricus, L rhamnosus* plus *L reuteri, B bifidum* plus *S thermophilus*, or *L rhamnosus* plus *L reuteri* were used in the childhood studies. In 20 studies reported between 1980 and 2004, all of the authors felt there was a definite benefit. Nineteen studies were either randomized with placebo, or randomized blinded with placebo, or placebo-controlled; one was an open trial [13–31].

Table 1
Probiotics used in adults with diarrhea

Author/reference	Type of study	Organisms	Dose	Result*
Sherakova and colleagues [6]	Randomized placebo	*L reuteri*	2.5 × 10 [7]	Shortened duration
Mansoon and colleagues [7]	Randomized double blind	*S boulardii*	250 mg TID	Shortened duration
Guslandi and colleagues [8]	Clinical trial	*S boulardii*	250 mg TID	Most remitted
Roza and colleagues [9]	Triple blinded placebo	*L GG*	1×10^{10} BID	Reduced frequency
Agorouaz and colleagues [10]	Randomized double blind	*L casei*	2×10^{8} TID	Shortened duration
Pelone and colleagues [11]	Randomized double blind placebo	*L casei/ S thermophilus/ B bulgaris*	5×10^{9}	Shortened duration
Xiao and colleagues [12]	Randomized placebo	*L acidophilus*	BID or TID	Partial improvement

Abbreviations: BID, two times daily; TID, three times daily.
*Result recorded as interpretation of data by authors.

Huang and colleagues performed a meta-analysis on children who were on standard rehydration therapy plus a probiotic and less than 5 years old [32]. These data stood up when the authors reanalyzed the material and limited it to hospitalized children in double-blinded studies employing *Lactobacillus*. Their meta-analysis of 18 published random double-blind controlled studies published between 1977 and 2001 clearly revealed that the duration of acute infection was decreased by 1 day when a probiotic was begun at onset of diarrhea. The authors' analysis also indicates that probiotic therapy decreases the length and severity of an acute attack of infectious diarrhea.

Antibiotic-associated diarrhea

Table 3 lists the 13 studies that were reported in the literature since 1980 that employed probiotics in an attempt to control antibiotic-associated diarrhea [18,26,33–43]. All of these studies were controlled either by randomization, blind, or placebo, or a combination, except for one open trial. Eight of the authors reported a definite prevention of antibiotic-associated diarrhea, whereas five reported negative results.

In a careful meta-analysis, DeSouza and colleagues [44] chose nine randomized, double-blind placebo-controlled trials for evaluation. Their attempt simply pointed out that a meta-analysis is difficult to do in this group, because many organisms were employed such as *L acidophilus*, *L casei*, *L bulgaricus*, *Bifidobacterium*, *Enterococcus faecium*, and *S boulardii*. DeSouza and colleagues analyzed results in 1214 patients, and the pooled data showed that probiotics were more effective than placebo. Ninety percent of the patients in the probiotic group did not have diarrhea, compared with 78% of the patients

Table 2
Probiotics used in children with diarrhea

Author/reference	Type of study	Organisms	Dose	Result**
Simakachorn and colleagues [13]	Randomized double-blinded placebo	*L acidophilus*	1×10^{10}	Decreased duration
Pedone and colleagues [14]	Randomized double-blinded placebo	*L casei*	*	Decreased severity
Shornikova and colleagues [15]	Randomized double-blinded placebo	*L reuteri*	1×10^{11}	Reduced diarrhea
Guandalini and colleagues [16]	Randomized double-blinded placebo	*L GG*	1.5×10^{10}	Shorter duration, rehydration course, hospital stay
Kaila and colleagues [17]	Randomized double-blinded	*L GG*	1×10^{11}	Reduced diarrhea
Vanderhoof and colleagues [18]	Randomized double-blinded placebo	*L GG*	2×10^{10}	Reduced frequency, increased stool consistency
Guarino and colleagues [19]	Randomized placebo	*L GG*	3×10^{9}	Reduced duration of rotavirus excretion
Majamaa and colleagues [20]	Randomized double-blinded	*L GG*	6.25×10^{9}	Reduced duration of diarrhea
Isolauri and colleagues [21]	Randomized placebo	*L GG*	1×10^{11}	Shorter duration
Arvola and colleagues [22]	Randomized placebo	*L GG*	2×10^{10}	Fewer incidence
Sudarmo and colleagues [23]	Randomized	*B bifidum*	*	Less frequent stools
Phuapradit and colleagues [24]	Randomized placebo	*B bifidum*	*	Protects against rotavirus infection
Lee and colleagues [25]	Randomized placebo	*L acidophilus/ B infantis*	1×10^{9}	Less frequency
Hoyos and colleagues [26]	Open trial	*L acidophilus/ B infantis*	2.5×10^{8}	Reduced incidence of NEC
Agarwal and colleagues [27]	Randomized double-blinded placebo	*Lactococcus lactis/ Leuconostac lactis*	*	Reduced duration
Agarwal and colleagues [27]	Randomized double-blinded placebo	*L casei/ L bulgaricus/ S thermophilus*	*	Reduced duration
Gaon and colleagues [28]	Randomized double-blinded placebo	*L casei/ L acidophilus*	1.75×10^{14}	Reduced diarrhea and vomiting

Table 2
(continued)

Author/reference	Type of study	Organisms	Dose	Result**
Rosenfeldt and colleagues [29]	Randomized double-blinded placebo	*L rhamnosus/ L reuteri*	1×10^{10}	Reduced diarrhea, rotavirus excretion
Saavedra and colleagues [30]	Randomized double-blinded placebo	*B bifidum/ S thermophilus*	*	Less incidence
Rosenfeldt and colleagues [31]	Randomized double-blinded placebo	*L rhamnosus/ L reuteri*	1×10^{10}	Reduced duration

*Dose not clear.
**Result as interpreted by author.

in the placebo group. Six of the nine trials showed a significant benefit of probiotics. Of the 13 studies the authors analyzed, eight of the authors clearly felt there was a benefit from using the probiotics. This is reasonable therapy, because the probiotic is employed at the time of onset of the use of the antibiotics and continues only until the time that the antibiotic therapy is terminated or for a short period following the use of the antibiotic. Because there is little harm from using a probiotic, it would seem that this therapy is important, significantly effective, and cost-effective. But, it remains unclear which probiotic and what dose are most effective.

Although most antibiotic-associated diarrhea is mild, pseudomembranous colitis may develop. This often is associated with *Clostridia difficile*. The antibiotic treatment of this type of colitis is usually effective, but at times *C difficile* becomes resistant to treatment, and because it is a spore-forming organism, reinfections are common. In some individuals, control of the re-infection becomes a difficult clinical problem. For that reason, probiotics have been employed, and *S boulardii* has been used in a successful trial to prevent reinfection. In a prospective, double-blind, control study of 193 patients, Surawicz and colleagues [37] found that the probiotic given at least 72 hours after start of the antibiotic and continued for 3 days after the antibiotic was stopped reduced the incidence of reinfection down to 7.2% from 14.6%. The efficacy for preventing antibiotic-associated diarrhea was 51%. Similarly, it was found that using *LGG* may be effective in reducing the relapses of *C difficile*, but many of the reports with this organism have varied results and are uncontrolled [18,22,39,40,44–46].

Radiation-induced diarrhea

Three studies in the literature report benefit of using probiotics while patients receive radiation for malignancy, or following radiation if diarrhea has occurred [47–49]. All improved the patient status, and the authors felt the probiotics were warranted and successful in decreasing radiation diarrhea. One

Table 3
Probiotics used in antibiotic-associated diarrhea

Author/reference	Type of study	Organisms	Dose	Result**
Wullt and colleagues [33]	Randomized double-blinded placebo	*L plantarum*	5×10^{10}	No conclusion
Lewis and colleagues [34]	Randomized double-blinded placebo	*S boulardi*	113 g	No effect
Bleichner and colleagues [35]	Randomized double-blinded placebo	*S boulardi*	0.5 g	Reduced diarrhea
McFarland and colleagues [36]	Randomized double-blinded placebo	*S boulardi*	1 g	Reduced diarrhea
McFarland and colleagues [37]	Randomized double-blinded placebo	*S boulardi*	1.5×10^{10}	Reduced diarrhea
Surawicz and colleagues [38]	Randomized double-blinded placebo	*S boulardi*	0.5 g	Reduced diarrhea
Thomas and colleagues [39]	Randomized double-blinded placebo	*L GG*	2×10^{10}	No effect
Pochapin and colleagues [40]	Randomized placebo	*L GG*	*	Reduced diarrhea
Hoyos and colleagues [26]	Open trial	*B infantis/ L acidophilus*	2.5×10^{8}	Reduced diarrhea
Tankanow and colleagues [41]	Randomized double-blinded placebo	*L acidophilus/ L bulgaricus*	5×10^{8}	No effect
Clements and colleagues [42]	Randomized double-blinded placebo	*L acidophilus/ L bulgaricus*	2×10^{8}	No effect
Arvola and colleagues [22]	Randomized placebo	*L GG*	2×10^{10}	Fewer incidence
Vanderhoof and colleagues [18]	Randomized double-blinded placebo	*L GG*	2×10^{10}	Reduced frequency, increased stool consistency

*Dose not clear.
**Result as interpreted by author.

study employed yogurts that contained *L acidophilus,* and the dose was twice a day for 3 weeks [47]. A second study employed *LGG* in a dose given three times a day, and the third study used the combination of eight organisms in VSL#3 three times a day. The VSL study included 190 patients who were randomized to receive placebo or VLS (450 billion organisms/g) three times daily. Fifty-two of 95 patients on placebo developed diarrhea, whereas only 8 of 95 patients on VSL developed diarrhea ($P < .001$). Although this is a small number of patients, it appears promising and warrants further study. The mechanism of probiotic action may be through the immumodulating effect or by regulation of apoptosis [47].

Vaginosis and vaginitis

Bacterial vaginosis and yeast vaginitis are common urogenital problems. It is estimated that more than 300 million cases of urinary tract infections occur annually on a worldwide basis [50]. The normal vaginal flora in premenopausal women consists primarily of lactobacilli. It has been assumed that these flora are protective against infection in normal women. Many pathophysiologic factors, however, cause unstable vaginal flora that may result in infection.

There have been relatively few studies with the use of probiotics. A classic experiment performed by Hilton and colleagues [51], in which they fed *L acidophilus,* resulted in a significant decrease of infections. When a control group of women who received a placebo was compared with women who were treated with yogurt containing probiotics, those on the placebo had 36 infections during the almost 1-year trial period, whereas the women who ate the yogurt only had four infections. This experiment indicated the importance of orally fed lactobacilli in maintaining immunoprotection. Other studies using vaginal installation of lactobacilli also have demonstrated benefit. The authors' search of the literature revealed nine manuscripts on the subject. Table 4 lists those studies. Six studies using *L acidophilus* reported a decrease in infections or symptoms [51–56], whereas three were primarily physiologic studies. One showed *LGG* failed to colonize the vagina [57], whereas two showed the same organisms, in conjunction with *L fermentum,* maintained healthy flora and reduced pathogens [58,59].

This published literature is too small to draw any long-term or therapeutic conclusions. The overwhelming experience of these investigators, however, revealed a positive effect of *L acidophilus.* Well-controlled randomized studies are needed in this area.

Helicobacter pylori

Several observers have felt that the addition of probiotics, by enhancing immunoregulation, would aid in the suppression of *Helicobacter pylori* infestation. The authors identified eight manuscripts that related *H pylori* infection to probiotic therapy. Two of them were clinical trials, and six of them were randomized studies as listed in Table 5. *L gasseri* [60], *LGG,* [61] *L casei* [62], and *L johnsoni* [63–67] were employed, but the latter organism was used in five studies. All of the reports indicated there was either a suppression of *H pylori* growth or a histologic suppression of the gastritis. No study, however, identified eradication of the organism. Therefore, the theory that probiotic therapy would enhance the disappearance of *H pylori* does not gain any strength from this literature.

Table 4
Vaginosis and vaginitis

Author/ reference	Type of study	Organism	Route	Dose	Effect
Hilton and colleagues [51]	Randomized placebo controlled crossover	*L acidophilus*	Yogurt	QD for 6 months	Decreased infections
Williams and colleagues [52]	Randomized double-blind placebo	*L acidophilus*	Vaginal	Weekly	Decreased yeast infections
Parent and colleagues [53]	Randomized placebo	*L acidophilus*	Table	QD for 6 days	Reduced episodes of bacterial vaginosis
Hallen and colleagues [54]	Randomized double-blind placebo	*L acidophilus*	Capsule	QD for 6 days	Reduced episodes of bacterial vaginosis
Pattman and colleagues [55]	Clinical trial	*L acidophilus*	Vaccination	TID for 2 weeks	Reduced symptoms
Shaley and colleagues [56]	Randomized	*L acidophilus*	Yogurt	QD for 2 months	Reduced episodes of bacterial vaginosis
Colodner and colleagues [57]	Randomized	*Lactobacillus GG*		BID for 4 weeks	Failed to colonize
Reid [58]	Randomized double-blind	*Lactobacillus rhamnosus/ L fermentation*	Capsule	QD for 4 weeks	Maintains healthy flora
Reid [59]	Randomized double-blind placebo	*L rhamnosus/ L fermentation*	Capsule	1×10^9 daily for 2 months	Reduced pathogenes

Abbreviations: QD, daily dose; BID, twice daily; TID, three times daily.

In another review article [68], six human trials were identified studying 180 subjects. Again, this review revealed some benefit but did not demonstrate any clear disappearance of *H pylori.*

The paper by Cremonini and colleagues [60] demonstrates the complexity of treating a widespread endemic infection like *H pylori* with probiotics. These authors identified 86 *H pylori*-positive asymptomatic patients and randomized them into four groups to receive a probiotic or placebo during and for 4 days after 1 week of triple therapy, which included rabeprazole, clarithromycin, and tinadizole. Another group received a combination of lactobacilli and bifidobacteria, and a fourth and final group only received a placebo. Although adverse effects occurred during the eradication week, all patients completed therapy. In the patients receiving probiotics, there was significantly less

Table 5
Helicobacter pylori

Author/ reference	Type of study	Organism	Route	Dose	Effect
Cremonini and colleagues [60]	Randomized double-blind placebo	*Lactobacillus GG*	Yogurt	QD for 8 weeks	Suppressed *H pylori*, decreased inflammation
Sakamoto and colleagues [61]	Clinical trial	*L gasseri*	Sachet	BID for 2 weeks	Decreased adverse effects of antibiotic
Cruchet and colleagues [62]	Randomized double-blind placebo	*L johnsonii*	Milk	QD for 4 weeks	Reduced *H pylori* population
Felley and colleagues [63]	Randomized double-blind placebo	*L johnsonii*	Milk	BID for 3 weeks	Reduced *H pylori* population and inflammation
Michetti and colleagues [64]	Randomized double-blind placebo	*L johnsonii*	Whey supernatant	2 weeks	Reduced *H pylori* population and inflammation
Pantoflickova and colleagues [65]	Randomized double-blind placebo	*L johnsonii*	Milk	3 weeks	Decreased gastritis
Gotteland and colleagues [66]	Clinical trial	*L johnsonii*	Whey supernatant	BID for 2 weeks	Reduced *H pylori* population
Cats and colleagues [67]	Randomized	*L casei*	Milk	TID for 3 weeks	Reduced *H pylori*

Abbreviations: QD, daily dose; BID, twice daily; TID, three times daily.

diarrhea and taste disturbance during the therapy week compared with the placebo patients. The *H pylori* eradication rate, however, was identical between the probiotic and placebo groups. Therefore, the conclusion was that probiotics helped decrease adverse effects but that there was no significant lack of compliance and there was no clear difference in the effect of any of the probiotics used.

Although there is interest in the use of probiotics in the treatment of *H pylori* from the review of this literature, there is no clear pattern for which a recommendation can be made.

Immunologic modulation

The second largest group of studies identified concerned immunologic modulation [69–85]. There were 17 studies reported between 1980 and 2004

Table 6
Immunologic modulation

Author/ reference	Type of study	Organism	Route	Dose	Effect
Schiffrin and colleagues [69]	Randomized	*L acidophilus*	Milk	2×10^{10} TID 3 weeks	Increased phagocytosis
Spanhaak and colleagues [70]	Randomized	*L casei*	Milk	1×10^{11} TID 3 weeks	No effect
Cunningham-Rundles and colleagues [71]	Open trial	*L plantarum*	Liquid	1×10^{9} QD	Strong response
Cukrowska and colleagues [72]	Randomized blinded placebo	*E coli (Nissle)*	Liquid	1×10^{8} QD 3 weeks	Stimulates humoral cellular immunity
Gill and colleagues [73]	Randomized	*B lactis*	Milk	5×10^{10} QD 3 weeks	Increased cellular immunity
Arunachalam and colleagues [74]	Randomized double-blinded placebo	*B lactis*	Milk	3×10^{11} BID 6 weeks	Increased IFN alpha and phagocytosis
Chiang and colleagues [75]	Randomized double-blinded placebo	*B lactis*	Milk	2.5×10^{10} BID 9 weeks	Increased phagocytosis
Gill and colleagues [76]	Placebo	*B lactis*	Milk	5×10^{9} QD 3 weeks	Increased CD59 lymphs
Isolauri and colleagues [77]	Trial	*Lactobacillus GG*	Water	5×10^{10} QD 8 days	Increased Anti rotavirus IgM
Majamaa and colleagues [20]	Randomized double-blinded	*Lactobacillus GG*	Water	6×10^{9} BID 5 days	Increased IgA
Kaili and colleagues [78]	Randomized double-blinded	*Lactobacillus GG*	Water	1×10^{10} QD 5 days	Increased IgA
Pessi and colleagues [79]	Clinical trial	*Lactobacillus GG*	Capsule	1×10^{10} BID 4 weeks	Increased IL-10
Schiffrin and colleagues [80]	Randomized	*L johnsonii*	Milk	7×10^{10} QD 12 weeks	Increased phagocytosis
Marteau and colleagues [81]	Randomized double-blinded placebo	*L johnsonii*	Milk	1.5×10^{10} BID 4 weeks	Increased IgA

Table 6
(continued)

Author/ reference	Type of study	Organism	Route	Dose	Effect
Donnett-Hughes and colleagues [82]	Randomized double-blinded placebo	*S thermophilus*	Milk	1×10^8 QD 5 days	No effect
Sheih and colleagues [83]	Randomized double-blinded	*L rhamnosus*	Milk	2.5×10^{10} BID 3 weeks	Increased cellular response
DeSimone and colleagues [84]	Randomized placebo	*B bifidum* and *L acidophilus*	Capsule	2×10^9 QD 4 weeks	Increased blood B cell, decreased colon inflammatory response

Abbreviations: QD, daily dose; BID, twice daily; TID, three times daily.

(Table 6). Thirteen were randomized controlled; three were clinical trials, and one was placebo-controlled. Nine organisms were employed in the 13 studies, and one used two organisms. They were administered mostly in milk or liquid supplements, and the effects were noted with a range of doses, from 5×10^6 in the classic experiments of Isolauri and colleagues [77] on demonstrating increased antirotavirus IgM to much higher doses given anywhere from daily to four times a day over 5 days to 12 weeks. Fifteen of the 17 studies reported a definite immune response (see Table 6). The clearly demonstrated increased phagocytosis, a response by IgA and IgM, and in very selected experiments, B cell responses. Other cytokines such as alpha IFN and IL-10 also were stimulated by probiotic administration.

Evaluating this literature shows that there is a defuse and marked humoral and cellular response from taking probiotics.

As in all of the studies with probiotics, what remains to be determined is a clear guideline as to which probiotics should be used for which immune response. When attempting to induce optimal immunologic modulation, it may be wise to use multiple organisms, thereby stimulating a broad immune benefit. In vitro and animal experiments have not been included in this analysis. Nevertheless, the studies on a molecular level in animals and in vitro strengthen and fortify the observation that probiotic organisms stimulate and enhance the immune response [1,85].

INFLAMMATORY BOWEL DISEASE

Basic studies reveal that there are increased numbers of bacteria that may be more invasive in the mucous layer in patients with Crohn's disease, and there is a great biodiversity of the flora in patients who have Crohn's disease and

Table 7
Inflammatory bowel disease

Author/ reference	Type of study	Organism	Route	Dose	Effect
Ulcerative colitis					
Rembacken [86]	Randomized double dummy	*E coli (Nissle)*	Water	5×10^{10} BID	Same as mesalamine
Kruis [87]	Randomized double-blinded dummy	*E coli (Nissle)*	Capsule	5×10^{11} QD 12 weeks	Same as mesalamine
Ishikawa [88]	Randomized placebo	*Bifidobacteria* and *L acidophilus*	Liquid	1×10^{10} QD 1 year	Longer remission
Venturi [89]	Open trial	VSL#3*	Powder	1.5×10^{12} BID 1 year	Prolonged remission
Crohn's disease					
Malchow [90]	Randomized double-blinded placebo	*E coli (Nissle)*	Capsule	5×10^{10} QD 1 year	Longer remission
Plein [91]	Randomized double-blinded placebo	*S boulardi*	Capsule	250 mg TID 2 weeks	Decreased bowel movements
Guslandi [92]	Randomized	*S boulardi*	Capsule	1000 mg QD 6 months	Less relapses
Gupta [93]	Randomized placebo	*Lactobacillus GG*	Tablet	1×10^{10} BID 6 months	Improved symptoms
Barclay [94]	Randomized double-blinded placebo	*S cerevisiae*	Capsule	500 mg 8 days/4 weeks	Disease activity increased
Prantera [95]	Randomized placebo	*Lactobacillus GG*	Powder	6×10^{9} BID 1 year	No effect
Pouchitis					
Kuisma [96]	Randomized double-blinded placebo	*Lactobacillus GG*	Capsule	2×10^{10} BID 3 months	No effect
Gionchetti [97]	Randomized double-blinded placebo	VSL#3*	Powder	3×10^{12} QD 9 months	Less relapses
Laake [98]	Open trial	*L acidophilus* and *Bifidobacteria*	Milk	5×10^{5} QD 4 weeks	Improved gross pathology

Table 7
(continued)

Author/ reference	Type of study	Organism	Route	Dose	Effect
Gionchetti [99]	Randomized double-blinded placebo	VSL#3*	Powder	9×10^{11} QD	Reduced relapses
Mimura [100]	Randomized placebo	VSL#3*	Powder	9×10^{11} QD 1 year	Longer remission

Abbreviations: QD, daily dose; BID, twice daily; TID, three times daily.

*VSL#3 = *Lactobacillus casei L plantarum, L acidophilus, L delbreueckii, Bifidobacterium longum, B breve, B infantis,* and *Streptococcus salivarius.*

patients who have ulcerative colitis. These facts have led to the support of probiotic use in these patients [2,3].

There are 15 reported studies in inflammatory bowel disease (IBD), four in ulcerative colitis, six in Crohn's disease and five in pouchitis [86–90]. The studies are tabulated in Table 7 and reveal that 12 of the 15 studies reported

Table 8
Irritable bowel syndrome

Author/ reference	Type of study	Organism	Route	Dose	Effect
Nobaek and colleagues [101]	Randomized double-blinded placebo	*L plantarum*	Drink	2×10^{10} QD 4 weeks	Decreased symptoms
Sen and colleagues [102]	Double-blinded placebo crossover	*L plantarum*	Oat gruel	6.5×10^{9} QD 4 weeks	No effect
Niedzielin and colleagues [103]	Randomized double-blinded placebo	*L plantarum*	Liquid	1×10^{10} BID 4 weeks	Decreased symptoms
O'Sullivan and colleagues [104]	Randomized double-blinded placebo crossover	*Lactobacillus GG*	Tablet	1×10^{10} QD	No effect
Kim and colleagues [105]	Randomized double-blinded placebo	VSL#3*	Powder	4.5×10^{11} QD 2 weeks	Less bloating

Abbreviations: QD, daily dose; BID, twice daily.

*VSL#3 = *Lactobacillus casei L plantarum, L acidophilus, L delbreueckii, Bifidobacterium longum, B breve, B infantis,* and *Streptococcus salivarius.*

Table 9
Effect on cardiovascular risk factors and blood lipids

Author/reference	Type of study	Organism	Route	Dose	Effect
Anderson and colleagues [106]	Randomized double-blinded placebo	*L acidophilus*	Yogurt	3×10^9 QD 3 weeks	Lowered cholesterol slightly
Schaafsma and colleagues [107]	Randomized double-blinded placebo crossover	*L acidophilus*	Yogurt	1×10^{10} TID	Lowered serum cholesterol
Naruszewicz and colleagues [108]	Randomized double-blinded placebo	*L plantarum*	Drink	2×10^{10} QD 6 weeks	Lowered cardiovascular risk factors
Bukowska and colleagues [109]	Randomized double-blinded placebo	*L plantarum*	Oatmeal	1×10^{10} QD 6 weeks	Lowered cholesterol
de Roos and colleagues [110]	Randomized placebo	*L acidophilus*			No effect
Bahijiri and colleagues [111]	Randomized double-blinded placebo crossover	*S cerevisiae*	Capsule	1000 mg QD 8 weeks	Increased high-density lipoprotein (HDL), lowered drops
		Brewer's yeast			
Li and colleagues [112]	Randomized double-blinded placebo	*Brewer's yeast*		10 g QD 12 months	Lowered triglycerides
Kiessling and colleagues [113]	Controlled trial	*L acidophilus*	Yogurt	3×10^{10} QD 2 weeks	Increased HDL
		B longum		3×10^7 QD 2 weeks	
Agerholm-Larsen and colleagues [114]	Randomized double-blinded placebo	*L acidophilus*	Yogurt	9×10^8 QD 8 weeks	Lowered blood pressure
		S thermophilus		4×10^9 QD 8 weeks	

Table 9
(continued)

Author/ reference	Type of study	Organism	Route	Dose	Effect
Kawase and colleagues [115]	Randomized placebo	*L casei*	Milk	4×10^{11} BID 8 weeks	Lowered blood pressure, triglycerides, and pathogenic index
		S thermophilus		1×10^{11} BID 8 weeks	
Richelsen and colleagues [116]	Randomized double-blinded placebo	*Enterococcus faecium*	Milk	2×10^{9} QD 8 weeks	Lowered low-density lipoprotein (LDL) and cholesterol
		S thermophilus		2×10^{11} QD 8 weeks	
Agerbaek and colleagues [117]	Randomized double-blinded placebo	*Enterococcus faecium*	Milk	4×10^{10} QD 6 weeks	Lowered LDL
		S thermophilus		1.4×10^{9} QD 6 weeks	
Xiao and colleagues [118]	Randomized placebo	*B longum*	Yogurt	5×10^{9} TID 4 weeks	Lowered cholesterol
		S thermophilus			

Abbreviations: QD, daily dose; BID, twice daily dose; TID, three times daily.

either improvement in the disease or a significant decrease in relapses and prolonged remissions.

The four studies in ulcerative colitis were performed with *Escherichia coli* (Nissle) [86,87], bifidobacteria plus *L acidophilus* [88], and VSL#3 [89]. All reported positive results.

In the six studies in Crohn's disease, *E coli* (Nissle) [90], *S boulardi* [91,92], and *LGG* [93] all reported prolonged remissions and decreased symptoms. However, one study with *S cerevisiae* [94], and one with *LGG* [95], however, revealed no significant effect. Overall in Crohn's disease, four of the studies revealed definite clinical improvement, and two were negative.

Pouchitis, an inflammation that occurs in an ileo anal anastomosis pouch after total colectomy, had five evaluations. One with *LGG* that revealed no effect [96]; one with *L acidophilus* and bifidobacteria showed improved gross pathology [98], and three done and reported with VSL#3 showed that the probiotic may prevent pouchitis, prolong remissions, and treat active disease [97,99,100].

Similar conclusions have been drawn by other authors [4]. When all data in the 15 studies are considered, probiotic therapy appears encouraging in IBD. Further long-term studies are awaited.

IRRITABLE BOWEL SYNDROME

There are numerous anecdotal reports on the use of probiotics in irritable bowel syndrome (IBS) [101–105]. Because IBS, as defined by the Rome II criteria, is a syndrome lasting for more than 3 months and may be either constipation or diarrhea associated with abdominal pain, multiple symptoms need to be evaluated in any study. Hence, it is very difficult to interpret the effect of treatment.

The literature reveals five randomized or blinded controlled studies (Table 8). Three of the five studies reported a definite decrease in symptoms. Three of them employed *L plantarum*, but the negative study that used this organism only had 12 patients, whereas the other two had many patients. The other negative study employed *LGG* and showed relatively no effect. A careful study using VSL#3 (eight organisms) in 25 patients was done. The patients had a 2-week run-in, then were randomized to the probiotic or placebo twice daily for 8 weeks. There was a significant decrease in bloating but no difference in transit time [105].

Because only three of the studies indicated a positive effect, and because the symptoms of IBS are so varied, it is difficult to draw any definite conclusions. Although there is a trend in some of these studies, one can draw no definite conclusion on the use of probiotics in IBS. Many clinicians will feel it is worth a try, but yet large controlled studies are needed to draw any definite conclusions.

METABOLIC EFFECTS

Effect on cardiovascular risk factors and blood lipids

Table 9 lists 13 controlled studies that were discovered. Twelve of the studies report a positive effect on improving cardiovascular risk factors, and one reported no effect.

It is difficult to analyze these studies, because numerous different probiotics were used, and different outcomes were evaluated. Nevertheless, three studies reported a drop in serum cholesterol, two a drop in low-density lipoproteins, two an increase in high-density lipoproteins, and two a drop in systolic blood pressure. There is no question that we cannot draw conclusions from this range of variable studies. There is a trend, however, to decreasing risk factors. More controlled long-term evaluation would be of interest, but no conclusions can be drawn at this time.

Lactose intolerance

Lactose mal-digestion is the most common disorder of intestinal carbohydrate digestion in people. Seventy-five percent of adults are intolerant on a worldwide

Table 10
Summary of literature analysis

	Clinical effectiveness	Organisms
Adult diarrhea	+++	*Lactobacillus reuteri, Lactobacillus GG, L casei, L acidophilus, S boulardi**
Childhood diarrhea	+++	*Lactobacillus acidophilus, L casei, L GG, Bifidobacteria**
Antibiotic diarrhea	+++	*L plantarum, S boulardi, L GG**
Radiation	++	VSL#3
Vaginosis	++	*Lactobacillus acidophilus*
H pylori	?	*L johnsonii, L GG, L gasseri, L casei*
Immunologic modulation	+++	*L acidophilus, L plantarum, B lactis, L GG, L johnsonii**
Ulcerative colitis	++	*E coli (Nissle), Bifidobacteria* and *Lactobacillus*, VSL#3
Crohn's disease	?	*E coli (Nissle), S Boulardi, L GG* (variable)
Pouchitis	+++	VSL#3
Irritable bowel syndrome	++	*L plantarum*, VSL#3
Cardiovascular risk	?	*L acidophilus, L plantarum**
Lactose intolerance	?	*L acidophilus, B longum**
Constipation	?	*L casei, L reuteri, L GG, B animalis, S cerevisiae*

VSL#3 = *Lactobacillus casei, L plantarum, L acidophilus, L delbreueckii, Bifidobacterium longum, B breve, B infantis*, and *Streptococcus salivarius*.
+++, strong evidence.
++, suggestive evidence.
?, too few studies to be certain.
*, plus multiple organisms.

basis [119–123]. There is great ethnic variation, in that the incidence is much lower in northern Europeans, and may run anywhere from 50% to 85% percent in African-Asians and anywhere from 90% to 100% in Asians. The authors identified only five controlled studies on lactose intolerance. Saltzman and colleagues [119] working in Boston performed a randomized controlled study where subjects ingested *L acidophilus* twice daily for 7 days with breath hydrogen tests performed at baseline and after the *Lactobacillus* ingestion. They found no effect from the probiotic. The four other studies, however, all reported decreased hydrogen production in subjects taking a probiotic. *L acidophilus, B longum*, and a combination of *L acidophilus* and bifidobacteria were employed in three of the studies. A study in which patients were randomized in a crossover design to receive 500 g daily of fresh yogurt containing *L acidophilus* and bifidobacteria or heated yogurt reported interesting outcomes [122]. They studied lactose-intolerant and nonlactose intolerant subjects and measured short-chain fatty acids and performed breath hydrogen tests. They concluded the consumption of yogurt containing live bacteria merely improves lactose malabsorption, but they also noted increased propionate concentrations. This study agreed with the other three, and it would seem that ingestion of probiotics improves fermentation of lactose and normalizes breath hydrogen.

Constipation

The authors identified five controlled studies that employed five different probiotics for treating constipation [124–128]. All reported an increase in the number of bowel movements or a decrease in transit time. With such a variety of probiotics used in all of the studies using varied protocols, one cannot draw any conclusion other than there is a need for another large controlled study.

SUMMARY

The results of the authors' analysis are summarized in Table 10. It appears that probiotics are effective in helping shorten the course of adult and childhood diarrhea, in preventing and treating antibiotic-associated diarrhea, and in effecting immunomodulation and treating pouchitis. There is less evidence, but some exists that is convincing to suggest that they may be helpful for treating radiation diarrhea, vaginosis, ulcerative colitis, and IBS.

There are numerous papers and studies, but there is not enough evidence in such conditions as *H pylori*, Crohn's disease, cardiovascular risk factors, lactose intolerance, and constipation to conclude that probiotics are definitely helpful for these clinical conditions.

The use of probiotics has become important, and with more controlled and larger studies, one can expect that they will prove to be effective for treating many diseases.

References

[1] Macfarlane GT, Cummings JH. Probiotics and prebiotics: can regulating the activities of intestinal bacteria benefit health? BMJ 1999;318:999–1003.

[2] Floch MH. Prebiotics, probiotics and dietary fiber. In: Buchman A, editor. Clinical nutrition: a guide for gastroenterologists. Thorofare (NJ): Slack Incorporated; 2005, in press.

[3] Floch MH, Hong-Curtiss J. Probiotics and functional foods in gastrointestinal disorders. Curr Gastroenterol Rep 2001;3:343–50.

[4] Hart AL, Stagg AJ, Kamm MA. Use of probiotics in the treatment of inflammatory bowel disease. J Clin Gastroenterol 2003;36:111–9.

[5] Montrose DA, Floch MH. Probiotics used in human studies. J Clin Gastroenterol, in press.

[6] Shornikova AV, Casas IA, Mykkanen H, et al. Bacteriotherapy with *Lactobacillus reuteri* in rotavirus gastroenteritis. Pediatr Infect Dis J 1997;16(12):1103–7.

[7] Mansour-Ghanaei F, Dehbashi N, Yazdanparast K, et al. Efficacy of *Saccharomyces boulardii* with antibiotics in acute amoebiasis. World J Gastroenterol 2003;9(8):1832–3.

[8] Guslandi M, Giollo P, Testoni PA. A pilot trial of *Saccharomyces boulardii* in ulcerative colitis. Eur J Gastroenterol Hepatol 2003;15(6):697–8.

[9] Raza S, Graham SM, Allen SJ, et al. Lactobacillus GG promotes recovery from acute nonbloody diarrhea in Pakistan. Pediatr Infect Dis J 1995;14(2):107–11.

[10] Agarwal KN, Bhasin SK, Faridi MM, et al. Lactobacillus casei in the control of acute diarrhea–a pilot study. Indian Pediatr 2001;38(8):905–10.

[11] Pedone CA, Arnaud CC, Postaire ER, et al. Multi-centric study of the effect of milk fermented by *Lactobacillus casei* on the incidence of diarrhoea. Int J Clin Pract 2000;54(9):568–71.
[12] Xiao SD, Zhang de Z, Lu H, et al. Multi-center, randomized, controlled trial of heat-killed *Lactobacillus acidophilus* LB in patients with chronic diarrhea. Adv Ther 2003;20(5): 253–60.
[13] Simakachorn N, Pichaipat V, Rithipornpaisarn P, et al. Clinical evaluation of the addition of lyophilized, heat-killed *Lactobacillus acidophilus* LB to oral rehydration therapy in the treatment of acute diarrhea in children. J Pediatr Gastroenterol Nutr 2000 Jan;30(1): 68–72.
[14] Pedone CA, Bernabeu AO, Postaire ER, et al. The effect of supplementation with milk fermented by *Lactobacillus casei* (strain DN-114 001) on acute diarrhoea in children attending day care centres. Int J Clin Pract 1999;53(3):179–84.
[15] Shornikova AV, Casas I, Isolauri E, et al. *Lactobacillus reuteri* as a therapeutic agent in acute diarrhea in young children. J Pediatr Gastroenterol Nutr 1997;24:399–404.
[16] Guandalini S, Pensebene L, Zikri MA, et al. *Lactobacillus* GG administered in oral rehydration solution to children with acute diarrhea: a multi-center European trial. J Pediatr Gastroenterol Nutr 2000;30:54–60.
[17] Kaila M, Isolauri E, Saxelin M, et al. Viable versus inactivated *Lactobacillus* GG in acute rotavirus diarrhea. Arch Dis Child 1995;72:51–3.
[18] Vanderhoof JA, Whitney DB, Antonson DL, et al. *Lactobacillus* GG in the prevention of antibiotic associated diarrhea in children. J Pediatr 1999;135(5):564–8.
[19] Guarino A, Canani RB, Spagnuolo MI, et al. Oral bacterial therapy reduces the duration of symptoms and of viral excretion in children with mild diarrhea. J Pediatr Gastroenterol Nutr 1997;25(5):516–9.
[20] Majamaa H, Isolauri E, Saxelin M, et al. Lactic acid bacteria in the treatment of acute rotavirus gastroenteritis. J Pediatr Gastroenterol Nutr 1995;20(3):333–8.
[21] Isolauri E, Juntunen M, Rautanen T, et al. A human *Lactobacillus* strain (*Lactobacillus casei* sp strain GG) promotes recovery from acute diarrhea in children. Pediatrics 1991;88(1):90–7.
[22] Arvola T, Laiho K, Torkkeli S, et al. Prophylactic *Lactobacillus* GG reduces antibiotic-associated diarrhea in children with respiratory infections: a randomized study. Pediatrics 1999;104(5):e64.
[23] Sudarmo SM, Ranuh RG, Rochim A, Soeparto P. Management of infant diarrhea with high-lactose probiotic-containing formula. Southeast Asian J Trop Med Public Health 2003;34(4):845–8.
[24] Phuapradit P, Varavithya W, Vathanophas K, et al. Reduction of rotavirus infection in children receiving bifidobacteria-supplemented formula. J Med Assoc Thai 1999; 82(Suppl 1):S43–8.
[25] Lee MC, Lin LH, Hung KL, et al. Oral bacterial therapy promotes recovery from acute diarrhea in children. Acta Paediatr Taiwan 2001;42(5):301–5.
[26] Hoyos AB. Reduced incidence of necrotizing enterocolitis associated with enteral administration of *Lactobacillus acidophilus* and *Bifidobacterium infantis* to neonates in an intensive care unit. Int J Infect Dis 1999;3(4):197–202.
[27] Agarwal KN, Bhasin SK. Feasibility studies to control acute diarrhoea in children by feeding fermented milk preparations Actimel and Indian Dahi. Eur J Clin Nutr 2002;56(Suppl 4):S56–9.
[28] Gaon D, Garcia H, Winter L, et al. Effect of *Lactobacillus* strains and *Saccharomyces boulardii* on persistent diarrhea in children. Medicina (B Aires) 2003;63(4):293–8.
[29] Rosenfeldt V, Michaelsen KF, Jakobsen M, et al. Effect of probiotic *Lactobacillus* strains in young children hospitalized with acute diarrhea. Pediatr Infect Dis J 2002;21(5):411–6.
[30] Saavedra JM, Bauman NA, Oung I, et al. Feeding of *Bifidobacterium bifidum* and *Streptococcus thermophilus* to infants in hospital for prevention of diarrhoea and shedding of rotavirus. Lancet 1994;344(8929):1046–9.

[31] Rosenfeldt V, Michaelsen KF, Jakobsen M, et al. Effect of probiotic *Lactobacillus* strains on acute diarrhea in a cohort of nonhospitalized children attending day-care centers. Pediatr Infect Dis J 2002;21(5):417–9.
[32] Huang JS, Bousvaros A, Lee JW, et al. Efficacy of probiotic use in acute diarrhea in children. A meta-analysis. Dig Dis Sci 2002;47:2625–34.
[33] Wullt M, Hagslatt ML, Odenholt I. *Lactobacillus plantarum* 299v for the treatment of recurrent *Clostridium difficile*-associated diarrhoea: a double-blind, placebo-controlled trial. Scand J Infect Dis 2003;35:365–7.
[34] Lewis SJ, Potts LF, Barry RE. The lack of therapeutic effect of *Saccharomyces boulardii* in the prevention of antibiotic-related diarrhoea in elderly patients. J Infect 1998;36(2):171–4.
[35] Bleichner G, Blehaut H, Mentec H, et al. *Saccharomyces boulardii* prevents diarrhea in critically ill tube-fed patients. A multi-center, randomized, double-blind placebo-controlled trial. Intensive Care Med 1997;23(5):517–23.
[36] McFarland LV, Surawicz CM, Greenberg RN, et al. A randomized placebo-controlled trial of *Saccharomyces boulardii* in combination with standard antibiotics for *Clostridium difficile* disease. JAMA 1994;271(24):1913–8. Erratum in: JAMA1994;272(7):518.
[37] McFarland LV, Surawicz CM, Greenberg RN, et al. Prevention of beta-lactam-associated diarrhea by *Saccharomyces boulardii* compared with placebo. Am J Gastroenterol 1995;90(3):439–48.
[38] Surawicz CM, Elmer GW, Speelman P, et al. Prevention of antibiotic-associated diarrhea by *Saccharomyces boulardii*: a prospective study. Gastroenterology 1989;96(4):981–8.
[39] Thomas MR, Litin SC, Osmon DR, et al. Lack of effect of *Lactobacillus* GG on antibiotic-associated diarrhea: a randomized, placebo-controlled trial. Mayo Clin Proc 2001;76(9):883–9.
[40] Pochapin M. The effect of probiotics on *Clostridium difficile* diarrhea. Am J Gastroenterol 2000;95(Suppl 1):S11–3.
[41] Tankanow RM, Ross MB, Ertel IJ, et al. A double-blind, placebo-controlled study of the efficacy of Lactinex in the prophylaxis of amoxicillin-induced diarrhea. DICP 1990;24(4):382–4.
[42] Clements ML, Levine MM, Black RE, et al. *Lactobacillus* prophylaxis for diarrhea due to enterotoxigenic *Escherichia coli*. Antimicrob Agents Chemother 1981;20(1):104–8.
[43] D'Souza AL, Rajkumar C, Cooke J, Bulpitt CJ. Probiotics in prevention of antibiotic-associated diarrhoea: meta-analysis. BMJ 2002;324:1361.
[44] Gorbach SL, Chang TW, Goldin B. Successful treatment of relapsing *Clostridium difficile* colitis with *Lactobacillus* GG [letter]. Lancet 1987;2:1519.
[45] Biller JA, Katz AJ, Flores AF, et al. Treatment of recurrent *Clostridium difficile* colitis with *Lactobacillus* GG. J Pediatr Gastroenterol Nutr 1995;21:224–6.
[46] Bennett RG, Gorbach SL, Goldin BR, et al. Treatment of relapsing *Clostridium difficile* diarrhea with *Lactobacillus GG*. Nutr Today 1996;31(Suppl 1):35S–8S.
[47] Salminen E, Elomaa I, Minkkinen J, et al. Preservation of intestinal integrity during radiotherapy using live *Lactobacillus acidophilus* cultures. Clin Radiol 1988;39(4): 435–7.
[48] Urbancsek H, Kazar T, Mezes I, et al. Results of a double-blind, randomized study to evaluate the efficacy and safety of antibiophilus in patients with radiation-induced diarrhoea. Eur J Gastroenterol Hepatol 2001;13(4):391–6.
[49] Delia P, Sansotta G, Donato V, et al. Prevention of radiation-induced diarrhea with the use of VSL#3, a new high-potency probiotic preparation. Am J Gastroenterol 2002; 97(8):2150–2.
[50] Reid G. Probiotic agents to protect the urogenital tract against infection. Am J Clin Nutr 2001;73(suppl):437S–43S.
[51] Hilton E, Isenberg HD, Alperstein P, et al. Ingestion of yogurt containing *Lactobacillus acidophilus* as prophylaxis for candidal vaginitis. Ann Intern Med 1992;116:353–7.

[52] Williams AB, Yu C, Tashima K, et al. Evaluation of two self-care treatments for prevention of vaginal candidiasis in women with HIV. J Assoc Nurses AIDS Care 2001;12(4):51–7.
[53] Parent D, Bossens M, Bayot D, et al. Therapy of bacterial vaginosis using exogenously-applied *Lactobacilli acidophili* and a low dose of estriol: a placebo-controlled multi-centric clinical trial. Arzneimittelforschung 1996;46(1):68–73.
[54] Hallen A, Jarstrand C, Pahlson C. Treatment of bacterial vaginosis with lactobacilli. Sex Transm Dis 1992;19(3):146–8.
[55] Pattman RS, Sankar KN, Watson PG, et al. An audit of Gynatren (a *Lactobacillus acidophilus* lyophilisate) vaccination in women with recurrent bacterial vaginosis. Int J STD AIDS 1994;5(4):299.
[56] Shalev E, Battino S, Weiner E, et al. Ingestion of yogurt containing Lactobacillus acidophilus compared with pasteurized yogurt as prophylaxis for recurrent candidal vaginitis and bacterial vaginosis. Arch Fam Med 1996;5(10):593–6.
[57] Colodner R, Edelstein H, Chazan B, et al. Vaginal colonization by orally administered *Lactobacillus rhamnosus* GG. Isr Med Assoc J 2003;5(11):767–9.
[58] Reid G, Beuerman D, Heinemann C, et al. Probiotic Lactobacillus dose required to restore and maintain a normal vaginal flora. FEMS Immunol Med Microbiol 2001;32(1):37–41.
[59] Reid G, Charbonneau D, Erb J, et al. Oral use of Lactobacillus rhamnosus GR-1 and *L fermentum* RC-14 significantly alters vaginal flora: randomized, placebo controlled trial in 64 healthy women. FEMS Immunol Med Microbiol 2003;35(2):131–4.
[60] Sakamoto I, Igarashi M, Kimura K, et al. Suppressive effect of *Lactobacillus gasseri* OLL 2716 (LG21) on *Helicobacter pylori* infection in humans. J Antimicrob Chemother 2001;47(5):709–10.
[61] Cremonini F, Di Caro S, Covino M, et al. Effect of different probiotic preparations on anti-*Helicobacter pylori* therapy-related side effects: a parallel group, triple blind, placebo-controlled study. Am J Gastroenterol 2002;97(11):2744–9.
[62] Cats A, Kuipers EJ, Bosschaert MA, et al. Effect of frequent consumption of a *Lactobacillus casei*-containing milk drink in *Helicobacter pylori*-colonized subjects. Aliment Pharmacol Ther 2003;17(3):429–35.
[63] Cruchet S, Obregon MC, Salazar G, et al. Effect of the ingestion of a dietary product containing *Lactobacillus johnsonii* La1 on *Helicobacter pylori* colonization in children. Nutrition 2003;19(9):716–21.
[64] Felley CP, Corthesy-Theulaz I, Rivero JL, et al. Favourable effect of an acidified milk (LC-1) on *Helicobacter pylori* gastritis in man. Eur J Gastroenterol Hepatol 2001;13(1):25–9.
[65] Michetti P, Dorta G, Wiesel PH, et al. Effect of whey-based culture supernatant of *Lactobacillus acidophilus* (johnsonii) La1 on *Helicobacter pylori* infection in humans. Digestion 1999;60(3):203–9.
[66] Pantoflickova D, Corthesy-Theulaz I, Dorta G, et al. Favourable effect of regular intake of fermented milk containing *Lactobacillus johnsonii* on *Helicobacter pylori*-associated gastritis. Aliment Pharmacol Ther 2003;18(8):805–13.
[67] Gotteland M, Cruchet S. Suppressive effect of frequent ingestion of *Lactobacillus johnsonii* La1 on *Helicobacter pylori* colonization in asymptomatic volunteers. J Antimicrob Chemother 2003;51(5):1317–9.
[68] Hamilton-Miller JM. The role of probiotics in the treatment and prevention of *Helicobacter pylori* infection. Int J Antimicrob Agents 2003;22(4):360–6.
[69] Schiffrin EJ, Rochat F, Link-Amster H, et al. Immunomodulation of human blood cells following the ingestion of lactic acid bacteria. J Dairy Sci 1995;78(3):491–7.
[70] Spanhaak S, Havenaar R, Schaafsma G. The effect of consumption of milk fermented by *Lactobacillus casei* strain Shirota on the intestinal microflora and immune parameters in humans. Eur J Clin Nutr 1998;52(12):899–907.
[71] Cunningham-Rundles S, Ahrne S, Bengmark S, et al. Probiotics and immune response. Am J Gastroenterol 2000;95(Suppl 1):S22–5.

[72] Cukrowska B, Lodlnova-Zadnlkova R, Enders C, et al. Specific proliferative and antibody responses of premature infants to intestinal colonization with nonpathogenic probiotic *E coli strain* Nissle 1917. Scand J Immunol 2002;55(2):204–9.

[73] Gill HS, Rutherfurd KJ, Cross ML, et al. Enhancement of immunity in the elderly by dietary supplementation with the probiotic *Bifidobacterium lactis* HN019. Am J Clin Nutr 2001;74(6):833–9.

[74] Arunachalam K, Gill HS, Chandra RK. Enhancement of natural immune function by dietary consumption of *Bifidobacterium lactis* (HN019). Eur J Clin Nutr 2000;54(3):263–7.

[75] Chiang BL, Sheih YH, Wang LH, et al. Enhancing immunity by dietary consumption of a probiotic lactic acid bacterium (*Bifidobacterium lactis* HN019): optimization and definition of cellular immune responses. Eur J Clin Nutr 2000;54(11):849–55.

[76] Gill HS, Rutherfurd KJ, Cross ML. Dietary probiotic supplementation enhances natural killer cell activity in the elderly: an investigation of age-related immunological changes. J Clin Immunol 2001;21(4):264–71.

[77] Isolauri E, Joensuu J, Suomalainen H, et al. Improved immunogenicity of oral D × RRV reassortant rotavirus vaccine by *Lactobacillus casei* GG. Vaccine 1995;13(3):310–2.

[78] Kaila M, Isolauri E, Saxelin M, et al. Viable versus inactivated *Lactobacillus* strain GG in acute rotavirus diarrhoea. Arch Dis Child 1995;72(1):51–3.

[79] Pessi T, Sutas Y, Hurme M, et al. Interleukin-10 generation in atopic children following oral *Lactobacillus rhamnosus* GG. Clin Exp Allergy 2000;30(12):1804–8.

[80] Schiffrin EJ, Brassart D, Servin AL, et al. Immune modulation of blood leukocytes in humans by lactic acid bacteria: criteria for strain selection. Am J Clin Nutr 1997;66(2):515S–20S.

[81] Marteau P, Vaerman JP, Dehennin JP, et al. Effects of intrajejunal perfusion and chronic ingestion of *Lactobacillus johnsonii* strain La1 on serum concentrations and jejunal secretions of immunoglobulins and serum proteins in healthy humans. Gastroenterol Clin Biol 1997;21(4):293–8.

[82] Donnet-Hughes A, Rochat F, Serrant P, et al. Modulation of nonspecific mechanisms of defense by lactic acid bacteria effective dose. J Dairy Sci 1999;82(5):863–9.

[83] Sheih YH, Chiang BL, Wang LH, et al. Systemic immunity-enhancing effects in healthy subjects following dietary consumption of the lactic acid bacterium *Lactobacillus rhamnosus* HN001. J Am Coll Nutr 2001;20(Suppl 2):149–56.

[84] De Simone C, Ciardi A, Grassi A, et al. Effect of *Bifidobacterium bifidum* and *Lactobacillus acidophilus* on gut mucosa and peripheral blood B lymphocytes. Immunopharmacol Immunotoxicol 1992;14:331–40.

[85] Isolauri E, Sütas, Kankaanpää P, et al. Probiotics: effect on immunity. Am J Clin Nutr 2001;73:444S–50S.

[86] Rembacken BJ, Snelling AM, Hawkey PM, et al. Nonpathogenic *Escherichia coli* versus mesalazine for the treatment of ulcerative colitis: a randomized trial. Lancet 1999; 354:635–9.

[87] Kruis W, Schutz E, Fric P, et al. Double-blind comparison of an oral *Escherichia coli* preparation and mesalazine in maintaining remission of ulcerative colitis. Aliment Pharmacol Ther 1997;11(5):853–8.

[88] Ishikawa H, Akedo I, Umesaki Y, et al. Randomized controlled trial of the effect of bifidobacteria-fermented milk on ulcerative colitis. J Am Coll Nutr 2003;22(1):56–63.

[89] Venturi A, Gionchetti P, Rizzello F, et al. Impact on the composition of the faecal flora by a new probiotic preparation: preliminary data on maintenance treatment of patients with ulcerative colitis. Aliment Pharmacol Ther 1999;13(8):1103–8.

[90] Malchow HA. Crohn's disease and *Escherichia coli*. A new approach in therapy to maintain remission of colonic Crohn's disease? J Clin Gastroenterol 1997;25(4): 653–8.

[91] Plein K, Hotz J. Therapeutic effects of *Saccharomyces boulardii* on mild residual symptoms in a stable phase of Crohn's disease with special respect to chronic diarrhea—a pilot study. Z Gastroenterol 1993;31(2):129–34.

[92] Guslandi M, Mezzi G, Sorghi M, et al. *Saccharomyces boulardii* in maintenance treatment of Crohn's disease. Dig Dis Sci 2000;45:1462–4.
[93] Gupta P, Andrew H, Kirschner BS, et al. Is *Lactobacillus* GG helpful in children with Crohn's disease? Results of a preliminary, open-label study. J Pediatr Gastroenterol Nutr 2000;31(4):453–7.
[94] Barclay GR, McKenzie H, Pennington J, et al. The effect of dietary yeast on the activity of stable chronic Crohn's disease. Scand J Gastroenterol 1992;27(3):196–200.
[95] Prantera C, Scribano ML, Falasco G, et al. Ineffectiveness of probiotics in preventing recurrence after curative resection for Crohn's disease: a randomized controlled trial with *Lactobacillus* GG. Gut 2002;51:405–9.
[96] Kuisma J, Mentula S, Jarvinen H, et al. Effect of *Lactobacillus rhamnosus* GG on ileal pouch inflammation and microbial flora. Aliment Pharmacol Ther 2003;17(4):509–15.
[97] Gionchetti P, Rizzello F, Venturi A, et al. Oral bacteriotherapy as maintenance treatment in patients with chronic pouchitis: a double-blind, placebo-controlled trial. Gastroenterology 2000;119:305–9.
[98] Laake KO, Line PD, Aabakken L, et al. Assessment of mucosal inflammation and circulation in response to probiotics inpatients operated with ileal pouch anal anastomosis for ulcerative colitis. Scand J Gastroenterol 2003;38(4):409–14.
[99] Gionchetti P, Rizzello F, Helwig U, et al. Prophylaxis of pouchitis onset with probiotic therapy: a double-blind, placebo-controlled trial. Gastroenterology 2003;124(5):1202–9.
[100] Mimura T, Rizzello F, Helwig U, et al. Once-daily high-dose probiotic therapy (VSL#3) for maintaining remission in recurrent or refractory pouchitis. Gut 2004;53(1):108–14.
[101] Nobaek S, Johansson M-L, Molin G, et al. Alteration of intestinal microflora is associated with reduction in abdominal bloating and pain in patients with irritable bowel syndrome. Am J Gastroenterol 2000;95:1231–8.
[102] Sen S, Mullan MM, Parker TJ, et al. Effect of *Lactobacillus plantarum* 299v on colonic fermentation and symptoms of irritable bowel syndrome. Dig Dis Sci 2002;47(11):2615–20.
[103] Niedzielin K, Kordecki H, Birkenfeld B. A controlled, double-blind, randomized study on the efficacy of *Lactobacillus plantarum* 299V in patients with irritable bowel syndrome. Eur J Gastroenterol Hepatol 2001;13(10):1143–7.
[104] O'Sullivan MA, O'Morain CA. Bacterial supplementation in the irritable bowel syndrome. A randomized double-blind placebo-controlled crossover study. Dig Liver Dis 2000;32(4):294–301.
[105] Kim HJ, Camilleri M, McKinzie S, et al. A randomized controlled trial of a probiotic, VSL#3, on gut transit and symptoms in diarrhoea-predominant irritable bowel syndrome. Aliment Pharmacol Ther 2003;17(7):895–904.
[106] Anderson JW, Gilliland SE. Effect of fermented milk (yogurt) containing *Lactobacillus acidophilus* L1 on serum cholesterol in hypercholesterolemic humans. J Am Coll Nutr 1999;18(1):43–50.
[107] Schaafsma G, Meuling WJ, van Dokkum W, et al. Effects of a milk product, fermented by *Lactobacillus acidophilus* and with fructo–oligosaccharides added, on blood lipids in male volunteers. Eur J Clin Nutr 1998;52(6):436–40.
[108] Naruszewicz M, Johansson ML, Zapolska-Downar D, et al. Effect of *Lactobacillus plantarum* 299v on cardiovascular disease risk factors in smokers. Am J Clin Nutr 2002;76(6):1249–55.
[109] Bukowska H, Pieczul-Mroz J, Jastrzebska M, et al. Decrease in fibrinogen and LDL-cholesterol levels upon supplementation of diet with *Lactobacillus plantarum* in subjects with moderately elevated cholesterol. Atherosclerosis 1998;137(2):437–8.
[110] de Roos NM, Schouten G, Katan MB. Yoghurt enriched with *Lactobacillus acidophilus* does not lower blood lipids in healthy men and women with normal to borderline high serum cholesterol levels. Eur J Clin Nutr 1999;53(4):277–80.

[111] Bahijiri SM, Mira SA, Mufti AM, et al. The effects of inorganic chromium and brewer's yeast supplementation on glucose tolerance, serum lipids and drug dosage in individuals with type 2 diabetes. Saudi Med J 2000;21(9):831–7.
[112] Li YC. Effects of brewer's yeast on glucose tolerance and serum lipids in Chinese adults. Biol Trace Elem Res 1994;41(3):341–7.
[113] Kiessling G, Schneider J, Jahreis G. Long-term consumption of fermented dairy products over 6 months increases HDL cholesterol. Eur J Clin Nutr 2002;56(9):843–9.
[114] Agerholm-Larsen L, Raben A, Haulrik N, et al. Effect of 8-week intake of probiotic milk products on risk factors for cardiovascular diseases. Eur J Clin Nutr 2000;54(4):288–97.
[115] Kawase M, Hashimoto H, Hosoda M, et al. Effect of administration of fermented milk containing whey protein concentrate to rats and healthy men on serum lipids and blood pressure. J Dairy Sci 2000;83(2):255–63.
[116] Richelsen B, Kristensen K, Pedersen SB. Long-term (6 months) effect of a new fermented milk product on the level of plasma lipoproteins—a placebo-controlled and double-blind study. Eur J Clin Nutr 1996;50(12):811–5.
[117] Agerbaek M, Gerdes LU, Richelsen B. Hypocholesterolaemic effect of a new fermented milk product in healthy middle-aged men. Eur J Clin Nutr 1995;49(5):346–52.
[118] Xiao JZ, Kondo S, Takahashi N, et al. Effects of milk products fermented by *Bifidobacterium longum* on blood lipids in rats and healthy adult male volunteers. J Dairy Sci 2003;86(7):2452–61.
[119] Saltzman JR, Russell RM, Golner B, et al. A randomized trial of *Lactobacillus acidophilus* BG2FO4 to treat lactose intolerance. Am J Clin Nutr 1999;69(1):140–6.
[120] Mustapha A, Jiang T, Savaiano DA. Improvement of lactose digestion by humans following ingestion of unfermented acidophilus milk: influence of bile sensitivity, lactose transport, and acid tolerance of *Lactobacillus acidophilus*. J Dairy Sci 1997;80(8):1537–45.
[121] Jiang T, Mustapha A, Savaiano DA. Improvement of lactose digestion in humans by ingestion of unfermented milk containing *Bifidobacterium longum*. J Dairy Sci 1996;79(5):750–7.
[122] Vesa TH, Marteau P, Zidi S, et al. Digestion and tolerance of lactose from yogurt and different semisolid fermented dairy products containing *Lactobacillus acidophilus* and bifidobacteria in lactose maldigesters–is bacterial lactase important? Eur J Clin Nutr 1996;50(11):730–3.
[123] Rizkalla SW, Luo J, Kabir M, et al. Chronic consumption of fresh but not heated yogurt improves breath-hydrogen status and short-chain fatty acid profiles: a controlled study in healthy men with or without lactose maldigestion. Am J Clin Nutr 2000;72(6):1474–9.
[124] Koebnick C, Wagner I, Leitzmann P, et al. Probiotic beverage containing *Lactobacillus casei* Shirota improves gastrointestinal symptoms in patients with chronic constipation. Can J Gastroenterol 2003;17(11):655.
[125] Ouwehand AC, Lagstrom H, Suomalainen T, et al. Effect of probiotics on constipation, fecal azoreductase activity and fecal mucin content in the elderly. Ann Nutr Metab 2002;46:159–62.
[126] Ling WH, Hanninen O, Mykkanen H, et al. Colonization and fecal enzyme activities after oral Lactobacillus GG administration in elderly nursing home residents. Ann Nutr Metab 1992;36(3):162–6.
[127] Wenk R, Bertolino M, Ochoa J, et al. Laxative effects of fresh baker's yeast. J Pain Symptom Manage 2000;19(3):163–4.
[128] Marteau P, Cuillerier E, Meance S, et al. *Bifidobacterium animalis* strain DN-173 010 shortens the colonic transit time in healthy women: a double-blind, randomized, controlled study. Aliment Pharmacol Ther 2002;16(3):587–93.

Gastroenterol Clin N Am 34 (2005) 571–579

GASTROENTEROLOGY CLINICS
OF NORTH AMERICA

INDEX

A

B

Note: Page numbers of article titles are in **boldface** type.

0889-8553/05/$ – see front matter
doi:10.1016/S0889-8553(05)00067-1

gastro.theclinics.com

M

Changing Your Address?

Make sure your subscription changes too! When you notify us of your new address, you can help make our job easier by including an exact copy of your Clinics label number with your old address (see illustration below.) This number identifies you to our computer system and will speed the processing of your address change. Please be sure this label number accompanies your old address and your corrected address—you can send an old Clinics label with your number on it or just copy it exactly and send it to the address listed below.

We appreciate your help in our attempt to give you continuous coverage. Thank you.

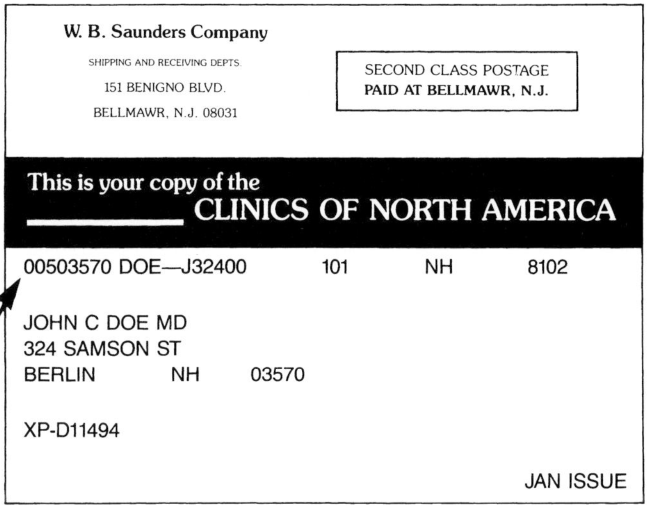
W. B. Saunders Company
SHIPPING AND RECEIVING DEPTS.
151 BENIGNO BLVD.
BELLMAWR, N.J. 08031

SECOND CLASS POSTAGE
PAID AT BELLMAWR, N.J.

This is your copy of the
__________ CLINICS OF NORTH AMERICA

00503570 DOE—J32400 101 NH 8102

JOHN C DOE MD
324 SAMSON ST
BERLIN NH 03570

XP-D11494

JAN ISSUE

Your Clinics Label Number

Copy it exactly or send your label
along with your address to:
W.B. Saunders Company, Customer Service
Orlando, FL 32887-4800
Call Toll Free 1-800-654-2452

Please allow four to six weeks for delivery of new subscriptions and for processing address changes.